"Geology is all about rocks, and rocks are all about detail and field context, and about actually being out there at that critical outcrop, right THERE, that proves our point. This new book is full of practical tips, commentary, and advice from real geologists who have been there and know what the science is all about. We are not, primarily, number crunchers and we are not, primarily, laboratory analysts. We are people who get mud on our boots and under our fingernails, people who understand space and time and scale and orientation as critical elements of an understanding of how the earth works. Don't let anybody persuade you otherwise — geology is a unique profession with a unique set of skills, and without us minerals and energy resources would not be found."

Andrew D Miall

Professor of Geology, University of Toronto

52 THINGS YOU SHOULD KNOW ABOUT GEOLOGY

EDITED BY MATT HALL

First published in 2014 by Agile Libre
Nova Scotia, Canada. *www.agilelibre.com*

Technical editor Matt Hall • *Managing editor* Kara Turner
Layout design Neil Meister, MeisterWorks • *Indexer* Linda Lefler
Series design electr0nika • *Cover design* Sari Naworynski

Library and Archives Canada Cataloguing in Publication

52 things you should know about geology / edited by Matt Hall
Includes bibliographical references and index.

ISBN 978-0-9879594-2-3 (pbk.)

1. Geology. I. Hall, Matt, 1971-, editor of compilation
II. Title: Fifty-two things you should know about geology.

QE26.3.F54 2013 550 C2013-906462-1

Who we are

Agile Libre is a small independent publisher of technical books in Nova Scotia, Canada. This is its second book, the first being *52 Things You Should Know About Geophysics*, published in 2012. Our passion is for sharing, so our books are openly licensed and inexpensive to buy.

Agile Libre is part of Agile Geoscience, a geoscience consulting company based in Nova Scotia, Canada. Find us at *agilegeoscience.com*.

Where to get this book

You will find this book for sale at *agilelibre.com*, and also at Amazon's various stores worldwide. Professors, chief geoscientists, managers, gift-givers: if you would like to buy more than 10 copies, please contact us for a discount at *hello@agilelibre.com*.

About open licenses

Colophon

This book was compiled in Microsoft Word, and laid out on a Mac using Adobe InDesign. The cover typeface is Avant Garde Gothic and the text typefaces are Minion and Myriad. The figures were prepared in Inkscape. It was published through Amazon's CreateSpace.

Contents

Alphabetical

Contents

By theme

Theme: FUNDAMENTALS

Tags: BASICS • MAPPING

INTEGRATED INTERPRETATION

INTEGRATION • INTERPRETATION

CALIBRATION

FIELDWORK • ANALYSIS

BASIN ANALYSIS

SED-STRAT • TECTONICS • REGIONAL

EXPLORATION

QUANTITATIVE

QUANTITATIVE • UNCERTAINTY • MODELLING

PROFESSION

CAREER • COMMUNICATION

SKILLS

METHODS • NINJA SKILLS

CULTURE

STORIES • TRAVEL • COMMENT

Introduction

Welcome to what is probably definitely the world's most readable book about petroleum geoscience. But it's not just for geologists — this book is for geophysicists, reservoir engineers, and exploration managers, too. It's for anyone that wants to think differently about the subsurface.

Between them, the 42 authors of this book have 850 years of experience as geoscientists. We can all learn from them, as we can from everyone in our profession. Sometimes it's hard to slow down enough to find a mentor, ask someone's opinion, or share a story of our own. But it's crucial for our science, based as it is on creative insight and evidence-based storytelling.

There is lots to chew on here. Despite their shortness, many of the essays cut to the core of earth science, asking what it is to know something about the earth's history — something we can never completely know with any certainty. This contrasts with the first book in this series, *52 Things You Should Know About Geophysics* because, to generalize horribly, geophysical unknowns are often formally recognized approximations or assumptions. In geological models, on the other hand, there are far more question marks than variables. Several of the authors touch on these important ideas — look for the **MODELLING** and **UNCERTAINTY** tags in particular.

Another observation: geologists seem to be more verbose than geophysicists. We had to split four essays in half to fit them in the book, and substantial cuts were made to 18 others. Brevity is usually all upside, but in some cases substantial material was cut. For example, you have to read all of Erik Lundin's wonderful smackdown of the one-size-fits-all mantle plume theory to really appreciate it.

These books challenge the expert culture we have in petroleum geoscience, and arguably in all of science and scientific publishing. Instead, they promote a culture of expertise, which is an entirely different thing. The idea is that we all have things we know or understand better than most, and at the very least have stories to share. So, while you will find essays here by much-lauded geoscientists — look for Mark Myers, S George Pemberton, Richard Hardman, Tony Doré, and Bruce Hart, for example — you will also find pieces by those at the start of their careers, such as Nicholas Holgate, Aruna Mannie, Tannis McCartney, Jesper Dramsch, and Zane Jobe. (An aside: three of those are acquaintances from the social media network Twitter, which is a wonderful source of lively conversation and even scientific inspiration for me.)

Another thing that sets this book apart from traditional technical publications is its open license. Copyright and licensing can seem complicated, so here's the nutshell version: the authors own their copyright, but have granted the world the right to use what they've written without permission, but with the requirement of attribution. The important point here, often misunderstood, is that the work is still copyright (that is, not public domain). The difference from, say, a typical book, is that you already have permission to use it for anything you like — you just have to give the author credit. I'd really like it if you mentioned the book too, but it's not a requirement.

The open license is intended to make the *Things You Should Know* easier to share. These little books are as much about our community of practice as they are about geoscience. Written for the community, by the community. To reflect this spirit, we're giving $2 from every sale to the AAPG Foundation.

I hope you enjoy reading this book. And I hope it inspires you to write. It's the most satisfying way I know of to connect with your community, and with the past and future of our exciting discipline.

Matt Hall
Nova Scotia, November 2013

Advice for a prospective geologist

Mark Myers

Geologists commonly work at time and space scales that are inconceivable to those in most other professions. We work with incomplete datasets — the preserved geological record represents a very small percentage of the total history. We work with complex and dynamic processes that require specialization and yet simultaneously broad interdisciplinary knowledge to unravel. In short, no matter what our particular area of specialization we need to have both an amazing breadth and depth of knowledge to be successful no matter if we are educators, researchers, or applied practitioners.

To that end I have four recommendations:

1. Don't specialize too quickly, but capture as much breadth as you can as an undergraduate. The high cost of college is driving both students and academic institutions to look for ways to accelerate the education process so that a higher percentage of the student population can graduate within four years. This focus is understandable but creates tension between opportunities for elective courses and undergraduate research on the one hand and getting the degree on time on the other. As I reflect back, some of my elective courses in geology, other sciences, and philosophy actually had the biggest influence on my professional career.
2. Spend as much time in the field as you can studying both the preserved geological record and modern systems. I was fortunate enough to have been able do field geology during the first 25 years of my career, much of it in remote parts of Alaska. Detailed description of the outcrop required me to work in four dimensions on complex problems and to develop a systematic methodology for separating observational data from interpretation. A strong background in field geology will be valuable even if you spend most of your professional career at a seismic workstation, at a microscope or probe, or in a remote-sensing lab.
3. Learn how to function and communicate effectively across disciplines both within and outside of the geosciences. Interdisciplinary science is necessary to understand the geological processes that have shaped our planet, discover and produce energy and mineral resources, manage and protect water resources, unravel geological history, manage and preserve ecosystem

Never let your intellectual curiosity stagnate.

services, or understand, predict, and monitor natural hazards. One of the great challenges in interdisciplinary science is being able to translate methodology and vocabulary between fields. This requires a significant effort of time and face-to-face interaction. It often requires willingness to go outside the comfort zone of your own frame of reference. While difficult, the personal growth and the advancement of scientific understanding make it rewarding and worthwhile.

4. Never let your intellectual curiosity stagnate. Hopefully you are considering the profession of geology or working as a geologist for more than the economic benefits. The opportunities to learn never stop. For me one of the greatest benefits is the amazing platform the profession provides to expand your mind. If your job doesn't provide you with a significant intellectual challenge and exciting opportunity to grow you are probably in the wrong field.

As easy as 1D, 2D, 3D

Nicholas Holgate, Aruna Mannie, and Chris Jackson

Observations in everyday life are commonly collected and described in one or two dimensions. Furthermore, these observations typically use the observer as the point of reference (that is, an egocentric reference); the most famous resultant fallacy of which is the Ptolemaic system or geocentric model. It is therefore no surprise that geologists struggle with incorporating multiple dimensions and perspectives.

Spatial thinking is an invaluable tool in the geologist's arsenal. Our understanding of earth processes has been greatly improved by our ability to collect and connect spatially disparate datasets. For example, the classification of crystalline systems (the fourteen 3D Bravais lattices) helped us understand crystal structure. Marie Tharp's topographic map of the Atlantic Ocean floor, drawn from thousands of sonar measurements and published in 1957, identified the Mid-Atlantic Ridge and propelled the theories of plate tectonics. And finally the demonstration by Richard Dixon Oldham that the arrival times of P-waves, S-waves, and surface waves from earthquakes could be explained by a dense planetary core (Oldham 1906), thereby paving the way for modern seismology.

Perhaps our everyday applications of spatial reasoning will not be quite as groundbreaking, but they are essential to both pure and applied geology. And the subjects that benefit from improved spatial reasoning are as equally wide-ranging as those examples. For example, our understanding of deformation patterns is enhanced by our ability to conduct balanced cross-section restoration. Structural field measurements and stereographic projections allow us to create cross-sections of the subsurface, but spatial reasoning impacts how accurately we can bring together these 3D data and present everything in a 2D form. Finally, even relatively simple tasks, such as locating a position on a topographic map from landscape observations, require us to have strong spatial reasoning skills (How far away is the river from the wall? What approximate elevation above the valley bottom am I?). It is therefore essential for a geologist to develop their spatial reasoning skills. But how?

Contrary to common assumption, advanced spatial reasoning ability is not a natural, innate aptitude. It can be taught and practised. For example, numerous studies have shown that response time decreases when attempting to conduct mental rotation after training (e.g. Kaushall and Parsons 1981; Kail and Park 1990). Whilst there is debate as to whether training improves the ability or simply allows the subject to retrieve previous mental rotations from memory (Heil et al. 1998), the result is nonetheless an improvement in spatial reasoning.

The most obvious way in which spatial reasoning is honed in geology is through fieldmapping. Not only does the participant learn how to translate the 3D terrain around them to a position on a 2D topographic map, as discussed above, but they must use spatial parameters such as perspective and more complex spatial reasoning to create subsurface cross-sections. In the UK at least, this is a part of undergraduate geology courses in which students either sink or swim.

The sinkers needn't be disheartened by their apparent inability to think spatially. It is something that can, and must, be practised. Spatial thinking is essential, but it is a skill that can be improved. The web is home to numerous teaching and training techniques: for example, look up *ageo.co/156bWPS*. And give this puzzle a go: which of the cubes on the right can be created by folding the net shown on the left?

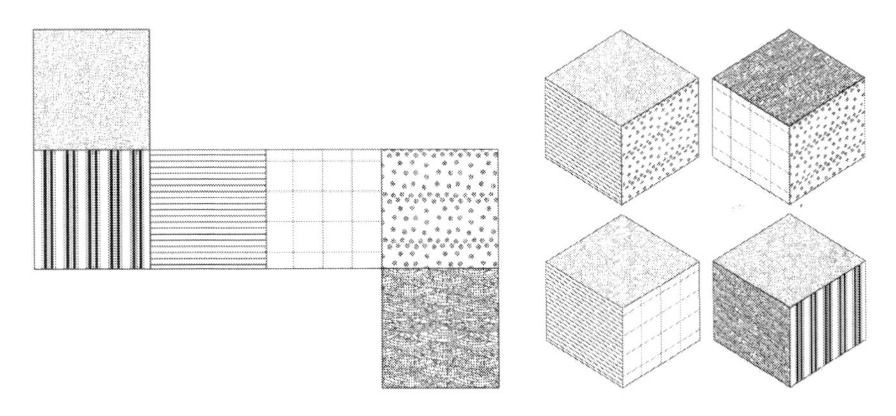

References

Heil, M, F Rösler, M Link, and J Bajric (1998). What is improved if a mental rotation task is repeated — the efficiency of memory access, or the speed of a transformation routine? *Psychological Research-psychologische Forschung* **61** (2), 99–106.

Kail, R and Y-S Park (1990). Impact of practice on speed of mental rotation. *Journal of Experimental Child Psychology* **49** (2), 227–244.

Kaushall, P and L M Parsons (1981). Optical information and practice in the discrimination of 3D mirror-reflected objects. *Perception* **10** (5), 545–562.

Oldham, R D (1906). The Constitution of the Interior of the Earth, as Revealed by Earthquakes. *Quarterly Journal of the Geological Society* **62**, 456–475.

Computational geology

Mark Dahl

A geologist is equal parts scientist and artist. If geology were solely science, the discipline would require nothing more than analysis of subsurface data at a series of control points. However, as there is no predictive power to understanding the control point only, a geoscientist must, using creativity and craftsmanship, model away from the known to approximate the subsurface across space and time. This is the essence of exploration.

Handcraft geology is a well-known concept amongst professionals within the discipline. Even if the term 'handcraft' is not commonly used, a strong handcraft culture exists. A corollary of the handcraft culture is a deep distrust of numerical techniques. For example, a hand-contoured map is always preferred to its numerically derived counterpart. Generally speaking, this culture is good for the discipline: while quantitative techniques cannot improve the fundamental ability of an interpreter, an inappropriately applied method can most certainly degrade an interpretation. I probably don't need to point out that the most advanced numerical tools have produced a great deal of poor interpretation. However, skillful use of numbers to interpret — what I call computational geology — will not erode the quality of a geologist's craft and will provide a number of opportunities not afforded to the handcraft geologist. Before getting into that, I need to explain more clearly what I mean by computational geology.

Computational geology is the deliberate use of numerical models to approximate an interpretation. The foundational interpretation remains the same as for the handcraft geologist: cores are described, stratigraphic tops picked, seismic horizons mapped, core data are analysed, and so on. The difference is the next step: instead of limiting himself to a sparsely distributed primary data source, and using heuristics to force a result, a computational geologist will look for his interpretation in the numbers. The computational geologist then builds a numerical model using data with superior spatial or temporal control that yields a result similar to the expected outcome. Here are a few examples. Rather than relying solely on pore pressure data, the computational geologist instead models a compaction trend from sonic logs or seismic data to characterize pore pressure in a basin. Similarly, rather than using only cored wells to characterize a reservoir, a computational geologist might model facies described in core using a linear algebraic equation of widely available wireline logs.

I have suggested that computational geology will not necessarily produce a superior result to the handcraft method — both techniques are ultimately limited by the interpreter's understanding of his craft — nevertheless there are a number of important advantages enjoyed by computational geology:

Iteration and validation. Handcraft is slow and tedious. A handcraft geologist typically produces a single deterministic model. There is often no time for a second interpretation. This severely limits a geologist's ability to characterize uncertainty and validate the model. Computational geology is a more efficient workflow where the human spends time on things only a human can do well (interpreting) and the computer takes care of the things the human is relatively inefficient and unskilled at (computation, repetition, precision). This leaves more time for iteration and validation with new data.

Uncertainty. Handcraft geology is not repeatable. When a handcraft geologist interpolates, she cannot quantify the likelihood of a contour being at a particular position and not another. Nor can she calculate the probability a facies boundary sits at a particular depth and not a metre lower. There is no rigorous way to quantify uncertainty for handcraft geology. In contrast, uncertainty quantification is intrinsic to all numerical models and is simply calculated as an attribute of the model. And while geologists love to criticize numeric interpolation, the numeric model ensures that all points in space are calculated consistently and, consequently, share a common uncertainty.

Integration. Handcraft geology tends to deal poorly with issues of spatial resolution. Detailed core description may not be calibrated to well logs. Core description is not commonly calibrated to seismic. This integration gap makes quantification very difficult. The work of a handcraft geologist is resolutely qualitative in nature and commonly integrates poorly with the quantitative subsurface disciplines — petrophysics, geophysics, basin modelling, and reservoir engineering. Consider that integration of the quantitative with qualitative aspects of subsurface characterization stands to benefit enormously from a numeric methodology, uniting the entire workflow.

Availability of digital data is rarely a limiting factor today. Creativity with respect to using these digits, however, is. Challenge yourself to make use of the advances in technology and make computational geology part of your workflow. You might be surprised at how well your interpretation can be teased out from the numbers.

Coping with uncertainty

Duncan Irving

The demands of today's geoscientist are many, not least in the pursuit of high quality geological insight. Mankind's interactions with our planet are often at scales that carry risk: infrastructure development, hydrocarbon and mineral extraction, and societal development in changing environments. Geological insights are consumed by end users far removed from the geoscientific tribe and these end users will have differing perspectives and perceptions of geoscientific output. The most fundamental question that can be asked of data and insights produced by a geologist is, 'how much can it be trusted?'

Trustworthy data allow geological models to be constrained and risks to be calculated. Trust in data is delivered by three components:

- The numerical error in a measurement, or the looseness of a subjective description. Is it 10 m? 10.00 m? 10.0 ± 0.1 m? The first is unhelpful, the second suggests a potential for high numerical precision with four significant figures, but only the third gives any feeling for how certain the user should be. Similarly describing samples from outcrop and core requires a high level of linguistic ability for it to be of use to anyone else in a time and place far removed from the initial inspection.
- The error in locating a measurement in space and time. With the advent of GPS and atomic clocks, locating something in space and time should be easy. It is trivial for a field team to keep a GPS logger switched on all day and apply waymarkers at each sampling location. Similarly all imagery collected should include information on the field of view, whether of a small outcrop or a complete vista. This allows simple integration of data into a virtual world or geospatial mash-up. Unfortunately, the best attempts are often thwarted by the diversity of coordinate reference systems, projections, and reference ellipsoids used in locating data. Any dataset or report providing spatial data should include the datum, reference ellipsoid, and coordinate system to give absolute confidence in location.
- Information and data about the measurement — often called metadata. This aspect of data quality is often unclear until you have need for it, but you will often end up wishing that more had been collected. It could be information about instrumentation such as calibration settings; it could be information

It is clear that the chain of data provenance is only as strong as its weakest link.

about the sampling environment — the weather, the sea state; it could be human information — who was on a field team; or it could be processing information — what version of which application was used to refine data or perform an interpretation.

These components are elements of a framework known as a data provenance architecture. There are organisations and groups (NASA, academic and government consortia, integrated oil companies) that perform geoscientific enquiry at massive scales and rates, and collect data from across many domains: geological, geophysical, oceanographic, engineering, even societal. Their workflows are complex and lengthy, and are typically revisited as new data become available for assimilation and re-evaluation. Data provenance architectures are emerging as a formal framework for the persistence of data quality and trustworthiness through such long and complex workflows. If implemented properly it should allow an end user to understand how and where the data were collected and, equally importantly, who has touched it and with what processes as it moves through a workflow. In the upstream oil industry many of the concepts are still aspirational but it is clear that the chain of data provenance is only as strong as its weakest link.

Current exemplary projects that address data provenance include:

- NASA's data management plan guidelines: *ageo.co/18H93wf*
- The USGS data management resources: *ageo.co/14rkIvR*
- The UK NERC data management plans: *ageo.co/1dwr6EL*
- UK scientific and computing academic communities: *ageo.co/16vtyGD* especially the ADMIRe program: *ageo.co/1aZZhXS*

Every geologist should be mindful of the part that they play in the chain and carry out their science with this in mind. They should strive to implement it in their daily work, as part of their corporate culture, and when procuring software and systems that facilitate their interaction with their data.

Geochemical alchemy

Richard Hardman

Alchemy was a medieval chemical philosophy aimed at turning base metals into gold. Long discredited, but ever hopeful, we humans continue with the fervently wished for in preference to the strictly logical. Explorers for oil and gas are always being approached with magical devices to help them. Many big, reputable companies have been conned this way. These days we apply geochemistry in much the same spirit, often pushing the conclusions beyond the bounds of logic.

In my career of over 50 years, many of the worst mistakes I've witnessed have come from too strict a consideration of geochemical principles. In the early stages of my career there was no clear accepted view of how oil and gas originates. I remember a talk in Colombia in 1966 that postulated that oil was of primary origin from the earth's crust and many years later in Sweden a deep well was drilled in Lower Palaeozoic rocks of the Siljan Ring to prove this idea.

One of the pioneers of oil industry geochemistry was Jim Momper who was based at Amoco's Research Laboratory in Tulsa. He was a great scientist and a real pioneer. When I joined Amoco in 1969, new hires were quickly indoctrinated. But like all 'holy writ,' eventually we started to realise that his ground-breaking work needed to be modified and expanded. For instance, he thought that carbonates could not be source rocks because they were not compressible. Saudi Arabia seemed to us to disprove this rather conclusively.

In 1975 Mesa Petroleum offered Amoco a farm-in to an Inner Moray Firth block they acquired in the fourth licensing round. We thought this was a no-hoper based on the shallow depth of burial of the source rocks. The Upper Jurassic crest of the structure is about 1800 m (6000 ft) deep, with the potential source rocks of the Middle and Lower Jurassic at about 2740 m (9000 ft). In 1976 the discovery well was drilled in block 11/30 and eventually the Beatrice Field with oil in place of 73 MSm^3 (460 million barrels) was found. There is still controversy about the source of all this oil, with the possibility that the Devonian may have contributed. At our own post-mortem the conclusion was that we had underestimated the geothermal gradient. The structure was formed as a result of movement along the Great Glen Fault, and mineral veins onshore — had we known about them at the time — would have given us the clue to hotter rocks. So strong was our adherence to geochemical principles I doubt whether

In exploration one should ... worry about the prospects which were evaluated and rejected, but which turned out to be big fields.

we could have persuaded the company to take a risk on what seemed a very risky venture. In exploration one should not worry about dry holes, they are an accepted hazard of the game. What one should really worry about are the prospects which were evaluated and rejected, but which turned out to be big fields.

On a more positive note, EnCore Oil led a partnership to the unlikely discovery of the Catcher Field in the central North Sea. This field and a cluster of others perhaps amounts to more than 24 MSm^3 (150 million barrels) of recoverable oil. In my days as head of exploration for Amerada Hess, we discovered the filled-to-spill Bittern Field in the Palaeocene, following which we and others searched up-dip for more prospects. Despite advanced seismic acquisition and processing, none were found. In 2010 EnCore decided to ignore conventional geochemical wisdom which said that any oil found even further up-dip in block 28/9 would be heavy and drilled 28/9-1. Oil was found at a depth of 1400 m (4600 ft) and, somewhat to the surprise of everyone, turned out to be crude of medium gravity. Thus ignoring geochemical advice led to a sound decision.

It seems geochemical calculations often are pushed beyond the limits of knowledge — either because a company wishes to do something bold or more likely because it does not. The important point is to be honest and avoid the presumption of understanding when none exists.

Geological inversion

Evan Bianco

Geological interpretation is the act of reasoning backwards. Geologists do what mathematicians and geophysicists call inverse problems all the time, but in their heads!

For instance, if I give you a rock, it is easy to determine its density, a property that may be of interest to you. All you need is a theory (gravity) and a method (say an Archimedes-like displacement of water). But if I give you density as a starting point, it is considerably harder — indeed, it's impossible — to reason backwards to reconstruct the rock from which it came. This is an inverse problem, and attempting to solve it is called inversion.

The difficulty of inverse problems is not due to some lack of knowledge about constituents and physical properties, but lies in all the different possible assemblies of grains, minerals, and fluids that could give the same result. Given such a lowly amount of data, we aren't adequately equipped to describe the rock any further without bringing in a priori information — the rock's appearance perhaps, or knowledge of where it was collected. Such problems are called under-determined: the number of unknowns exceeds the number of relationships or equations we can write down. We can't possibly solve them exactly, but it is still appealing to describe them approximately.

Forwards...

The opposite of an inverse problem is a forward problem. It uses straight-up reasoning or deduction starting with a model or theory and going towards a specific observation. The goal is to construct a test that can be compared with real data to confirm (or not) a theory you started with. Given this sedimentary environment, what sedimentary features do I expect to see? How will grain size vary laterally?

Another example: the classic forward model of a seismic trace links the unknowable earth (on the left in the figure) to a 1D record of lithology, to the speed of sound in those rocks, to acoustic reflectivity, and so to a seismogram (on the right). Seismic inversion is the inverse problem of deducing rocks from seismic.

Compared to interpretation these are easy problems, not because they are uncomplicated, but because there is a unique path between cause (model) and effect (data). Forward problems shouldn't be overlooked just because they are easy, however. Indeed, it is because they are solvable that we can use them to validate an interpretation or teach us what to look for.

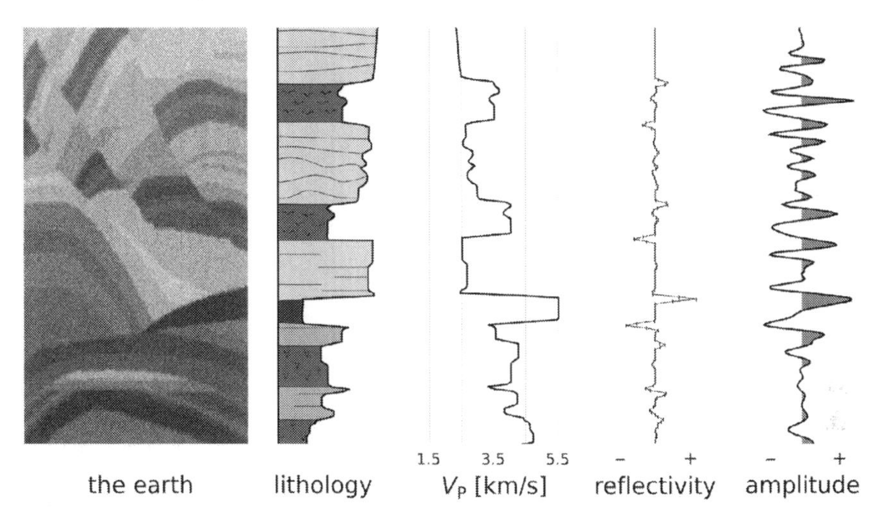

...and backwards

In this language, geological interpretation falls into the class of hard inverse problems — equivalent to moving to the left in the figure above. It begins with specific measurements, then we use inductive reasoning (informed guesses, basically) to detect patterns, regularities, and anomalies, and so formulate a hypothesis that we can explore. The output of a geological interpretation is a model full of assumptions. If the assumptions you brought to the problem are reasonable, the model can be considered meaningful. If you haven't considered and tested your assumptions, you haven't subscribed to reason.

Some are not aware, some choose to ignore, and some forget that works of geoscience are problems of extreme complexity. The only way we can cope with complexity is to make certain assumptions that make inverse problems solvable. If all you do is say, 'here is my interpretation,' you will be unconvincing. But if instead you ask, 'have I convinced you that my assumptions are reasonable?', it entirely changes the impact of your interpretation. It becomes a shared entity that anyone can look at and examine.

Acknowledgments

A version of this essay first appeared as a blog post in April 2013, *ageo.co/1elASMY*

Get a helicopter not a hammer

Alex Cullum

Science is built upon the ability to continually challenge, change, and refine our understanding of everything, but within a framework that respects what can be proven through controlled, repeatable, and documented analysis. Geological principles such as uniformitarianism, plate tectonics, evolution, and sequence stratigraphy are the result of detailed observation and measurement, but they have subsequently been proven and refined by scientific experiment with controls, constants, and mathematical theory. Geology's credentials as a science are solid, but there are issues which perhaps we could all take as 'areas for improvement' in the way we work, communicate, and conduct ourselves as geologists.

Geo-engineers working at the sharp end of geology — planning and drilling oil wells — often complain that geologists have too many arms. They claim that the geologist's answer to any emerging issue is always that 'on one hand it could be this, on the other hand it could be that, or it may be something else.' In other words, the geological model is constantly being updated with the latest data. In this sense geologists are being responsible scientists, but as an applied tool a constantly shifting base model can be the last thing that's required when all you want is help solving the current challenge.

Some economists question why companies drill exploration wells when the probability of a hydrocarbon discovery is on average less than 40 percent. Wouldn't the huge costs be better invested buying into existing secured volumes? With a bigger picture — a helicopter view — this is a flawed model since eventually you'll run out of existing volumes, but better communication of the risks and uncertainties in our work are certainly required at all levels.

Most geologists use their hammers (and brains) to dig deep pits of knowledge, sometimes referred to as academic silos. The deeper they get the more interesting things are for the individual and the harder it gets for the rest of us to relate to them. The more focused the individual gets, the narrower their field of view and the more restricted their overview of what the rest of us are up to. Discipline expertise is often what you need from someone, but more vital and harder to find is the competence and experience to frame and communicate a skill set within the bigger picture of the task at hand.

Geology's credentials as a science are solid, but there are issues which we could all take as 'areas for improvement' in the way we work, communicate, and conduct ourselves.

Focus on the level of information the investor, manager, co-worker, client, or audience actually requires. This bigger picture or helicopter view should be paramount otherwise the final outcome is likely to be that the work and its conclusions will be ignored, unused, or even discredited. If the recipient does not quickly and easily see the value in them, framed within the needs of their task, they will turn to other sources for support.

To achieve a helicopter view, the specific needs of the audience, customer, or colleague must be identified as early as possible.

- The timing and deadlines should be identified and the experience level of the audience or recipient should be assessed.
- How much time the audience or recipient has to absorb the results and conclusions should be determined.
- What does the customer need?
- Is a verbal confirmation within five minutes or a short practical presentation of facts or conclusions all that is needed?
- Perhaps a single page document or diagram should be emailed as the audience does not have time to read more. Many people appear to believe the product must always be a thesis of 200-plus pages in the most complex technical jargon possible; this is almost always *not* what is required.

Perhaps the most simple rule here is to ask yourself the control question, 'is this the product I want to deliver, or is it the delivery the audience/customer needed and asked for?' Even the most world-leading, groundbreaking geologist must put their ego to one side, assess the bigger context, and identify the needs of their audience if they wish to improve the communication of and application of their conclusions.

Get outside

Aruna Mannie, Nicholas Holgate, and Chris Jackson

The best geologist is the one who has seen the most rocks. This quote by Herbert Harold Read (1889–1970) may sound somewhat of a cliché. However, in a world of computerized technology where rock records are readily accessible via the Internet and analog databases, actually going out in the field and looking at outcrops or present-day geological processes is somewhat overlooked, but nevertheless important. Lathrop and Ebbett (2006) emphasized how essential field trips are in developing a proper sense of space, scale, and time. Fieldwork provides the training ground for understanding geological concepts and theories of how the earth works, which is critically important when we attempt to understand the 'hidden' subsurface.

Much of our understanding of the subsurface comes from interpreting seismic and well data. Too many times we are trapped in an office behind workstations interpreting seismic without stopping to ask what this seismic wiggle actually represents. Is it a geological feature that is the height of the Eiffel Tower? Does the reflector I'm mapping really represent a single geological time line? Are these faults actually discrete or are they associated with a broad damage zone? The ability to make this connection with the subsurface is improved by going

Geologists who spend time in the field... are able to see the geology in seismic data and fill the gaps between these and well data.

out in the field. It helps bridge the gap in scale and understanding between seismically resolvable features and real-world geology. This was emphasized by the self-evaluation survey results of Bentley (2009), who asked 'How do you learn geology best?' Before going on a field trip, 45 percent of those polled said that listening to formal lectures was best, whereas only 24 percent suggested that going on a field trip was the way to go. However, after the field trip the overall consensus was that field trips were equally as important as lectures for learning geology.

An accurate record of observations in the field in the form of notes, sketches, photographs, and maps, provides the foundation for developing ideas and testing different interpretations. Geology, unlike most other sciences, is a field science that does not provide a precise answer or a single interpretation. In fact, only by formulating different hypotheses and objectively evaluating them can we produce a sound, scientific basis for our interpretations. Our ability to do this is bolstered by spending time in the field. It's no surprise that geologists who spend time in the field, understanding the structural and stratigraphic relationships of rocks and geological processes, are those that are best able to see the geology in seismic data and fill the gaps between these and well data. Geology first originated in the field and, when we are stuck in understanding the subsurface, we venture back to field analogs — where it all started.

So, whether you are planning the next team-building event for the office, a trip for your local geological society, or an afternoon stroll on the beach, keep the rocks in mind.

References

Bentley, C (2009). *Touring and exploring: the role of field trips in geology education.* MSc thesis in Science of Education, Montana State University.

Lathrop, A and B Ebbett (2006). An inexpensive, concentrated field experience across the Cordillera. *Journal of Geoscience Education*, **54** (2), 165–171.

Get to know eigenvectors

Bernd Ruehlicke

Terms like orientation tensor (Scheidegger 1965), Bingham distribution (Bingham 1964), eigenvector and eigenvalue, symmetric matrix, and symmetry axis are probably long forgotten by you. They might fill you with fear, or curiosity. But what makes them important for a geologist, and how are they linked to locating more oil?

To see how, let's start with some synthetic data. An image log can help determine paleo-transport directions in a sedimentary system.

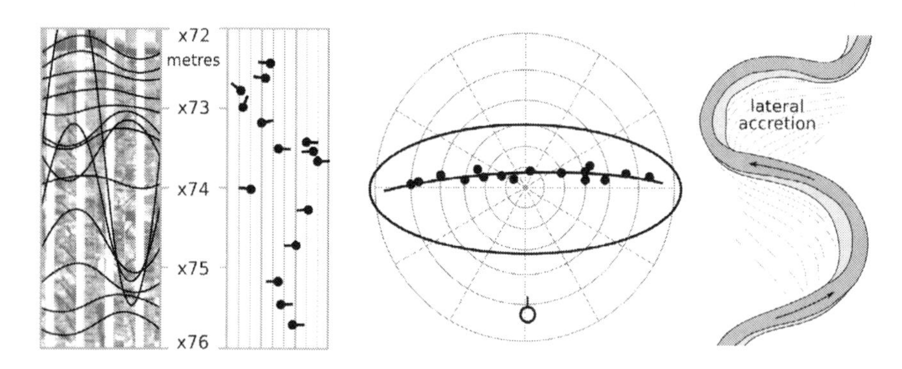

Planar features are manually interpreted and visualized on the image log as sine curves (above left). Often a 3D visualization of the planes on the image are used as guidance but this typically delivers more confusion than explanation. The identified planes are often represented as tadpoles on a depth-based plot, where the tadpoles correspond to the sine curves on the image log.

Stereographic projections of the planes are the primary tool to catch any possible symmetry of the (many) planar features, now abstracted as poles on a two dimensional plane (above middle). The open circle is the pole of the axis of the great circle which fits the poles of the planar features, and also the eigenvector corresponding to the the third eigenvalue of the orientation tensor based on their poles.

Poles are normal vectors to planes defined by polar dip and azimuth coordinates. The challenge now is to apply spherical statistics to differentiate between a cluster or girdle distribution (Woodcock 1977), and to find a corresponding measure of statistical significance. A cluster here is a set of poles with similar dip and azimuth, while poles on a girdle can be estimated by a great circle (very much like the flight path between Paris and New York) which has a plunge and a direction.

Linear algebra to the rescue

From linear algebra we know that a 3-by-3 symmetric matrix will have three eigenvectors with three corresponding eigenvalues. The eigenvalues, being real numbers, can be sorted in descending order, allowing us to refer to eigenvectors 1 to 3. For the full story, find 'Symmetric matrix' on Wikipedia.

It turns out that this is very helpful to us. The best mean pole of a population of poles is defined by the eigenvector belonging to the biggest eigenvalue of the 3-by-3 orientation tensor based on the given population. Even better, the third eigenvector (the one with the smallest eigenvalue with respect to the orientation tensor) defines the pole of the axis of the best great-circle fit through the given population of poles. This is useful!

Step by step

For simplicity, I have ignored structural dip removal for my synthetic example. The method would be:

1. Identify features on the image log, in this case an FMI image.
2. Look at the stereographic projection (forget 3D visualization — it will makes your eyes sick).
3. Perform spherical statistics by calculating the eigenvalues and eigenvectors of the orientation matrix based on the set of poles selected.
4. Incorporate other sources from the area, for example suggestions from seismic or other data.
5. *Voilà*! If we assumed a fluvial system or a channelized turbidite system, we have found a symmetry axis defined by the third eigenvector suggesting north–south trending channels (as shown in the figure).

References

Bingham, C (1964). *Distributions on the sphere and on the projective plane*, PhD dissertation, Yale University.

Scheidegger, A (1965). On the statistics of the orientation of bedding planes, grain axes, and similar sedimentological data. US Geological Survey Prof. Paper 525C.

Woodcock, N (1977). Specification of fabric shapes using an eigenvalue method, *Geological Society of America Bulletin* **88.**

Have hammer will travel

Dianne Tompkins

Geology, and petroleum geology in particular, could be called 'the creative science.' This is because while our interpretations may be rooted in science they are often brought to life by the imagination and creative talents of the individual interpreter. Petroleum geologists are not only creative in their work but are often inventive, resourceful, and adventurous in the way they live their lives.

I have lived and worked in many different places, and have visited countries I could only dream about growing up in Yorkshire, England. I was surrounded by geology from a young age — from the Millstone Grit of Ilkley Moors where we had our Girl Guide camp outs, to the Vale of York where glacial till masks the underlying bedrock. I grew up appreciating the wonders of nature and had no doubt I would somehow end up working among them.

I did not intend to become a petroleum geologist, however. I went to Aston University to study botany and agriculture but one semester of geology had me hooked. After graduating I continued my studies and focused on hard rock geochemistry, learning a lot of useful techniques such as electron microscopy and X-ray diffraction analysis that allowed me to later transfer to soft rock geology and enter the oil business. Finally I was a petroleum geologist, albeit an inexperienced one.

Have hammer will travel!

I served my apprenticeship, so to speak, and learned much about the industry in Aberdeen and offshore in the North Sea. Then the excitement really started! Two postings to China, working in a dynamic and exciting business and cultural environment, and eating creatures I had never before considered edible. Then on to the United States where a move to Odessa, Texas proved almost as great a culture shock as the move to China. Then eight years in Perth, Australia. What great geology, what prolific gas fields, and what a spectacular place to live. The Aussies are lovely people with their 'no worries' and 'she'll be right' culture.

Next, I moved back to Texas, to Houston, which is the heart and soul of the North American oil business. What a busy, crowded place! It was surely not Perth, but there were numerous opportunities and technical challenges there as we considered rocks that we once labelled unproductive 'tombstone' as

Always be open to change, new opportunities,
new places, and taking risks to follow your passion.

potential reservoirs. I am now living and working as a consultant geologist in Stavanger, Norway. So it seems that life comes full circle and I am again working in the North Sea almost 30 years after I started as an inexperienced but enthusiastic petroleum geologist in Aberdeen. It is lovely to be close to my family in Yorkshire while at the same time enjoying another country and culture so different to those I have already been fortunate enough to experience.

What next and would I do it again?

Petroleum geology is a creative science but also a traveller's science. Does all this travel make you a better geologist? Absolutely! If you subscribe to the adage, as I do, that the 'the best geologist is the one who has seen the most rocks' then it's clear you must take every chance to be adventurous.

Would I do anything differently if I had the chance? No, but if you're thinking of joining this industry you need to know that times are not always good and the nature of our work can be cyclical, in tune with economic booms and busts. The defence against this is to have a second string to your bow — develop transferable skills such as teaching or writing. Always be open to change, new opportunities, new places, and taking risks to follow your passion.

Finally, I think this quote succinctly describes what we do far better than I ever could:

> *No geologist worth anything is permanently bound to a desk or laboratory, but the charming notion that true science can only be based on unbiased observation of nature in the raw is mythology. Creative work, in geology and anywhere else, is interaction and synthesis: half-baked ideas from a bar room, rocks in the field, chains of thought from lonely walks, numbers squeezed from rocks in a laboratory, numbers from a calculator riveted to a desk, fancy equipment usually malfunctioning on expensive ships, cheap equipment in the human cranium, arguments before a road cut.*
>
> **Stephen Jay Gould**

References

Gould, S J (1987). *An Urchin in the Storm: Essays about Books and Ideas.* W W Norton and Company. 255 p.

Heterogeneity + sparse sampling = uncertainty

Michael Pyrcz and Clayton Deutsch

An important aspect of geology is the characterization of geological sites that are now frozen in time and space. The geological processes that led to the present-day preserved distribution of rock properties were acting at all scales and were transient, nonlinear, and chaotic; this leads to variability and heterogeneity at all scales. Our direct sampling of a geological site tends to be quite sparse because of drilling costs. Geophysical data are more extensive, but they include some noise in the measurements and measure properties at a larger scale than we must know them. It is axiomatic that heterogeneity and sparse sampling lead to uncertainty. Yet, there is but one inaccessible truth at the geological site.

This geological site with five wells has shale drapes that could have a large impact on production. The size, shape, and location of the shale drapes could never be predicted from the available well data; we are faced with significant uncertainty.

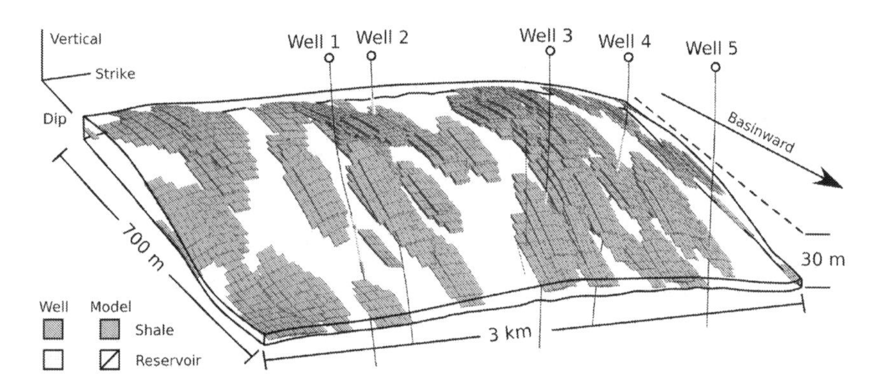

Decision makers want to quantify and manage uncertainty to make the best possible development decisions. They want decisions that are robust with respect to over- or under-predicting resources. They want to assess the value of new information. They want to compare the uncertainty between geological sites. The geologist must provide high-resolution 3D numerical models of the site that have realistic patterns of heterogeneity and, when taken together, provide a defensible assessment of uncertainty. Geostatistics provides tools for constructing such models given widely spaced measurements and various data types. Geostatistical tools have proven themselves in the last 25 years.

The application of geostatistical tools does not commonly fall within the conventional training of geologists. Most learn the tools from software vendors, short courses, and mentoring. There are few reference texts to help geologists with the complex interdependent modelling decisions that must be made; Pyrcz and Deutsch (2014) covers some important points. Here are three:

The physics of deposition, diagenesis, and structural deformation are not encoded in geostatistical algorithms. Geostatistical tools, including variogram-based methods, object-based methods, and multiple point statistics, permit the generation of numerical models that mimic our conceptual model within the limits of the inferred statistics. A well-reasoned conceptual model is essential. Often geostatistical models include a significant deterministic component informed directly by expert geological mapping. The results of geostatistical modelling may provide unexpected insights into the geological site through the integration of disparate data and concepts, but mostly the results are high-resolution numerical models for resource calculation and flow simulation.

Geostatistical sophistication cannot overcome bad input data and a flawed conceptual model. In fact the resulting heterogeneity models may mask data issues. We must accept the provided measurements and move on with numerical modelling, otherwise we will not have the decision support information that management requires. At the same time, we cannot simply take the data at face value and accept a preliminary and perhaps flawed conceptual model, otherwise the numerical models are not useful. Uncertainty and bias in the data and conceptual model, and alternative geological scenarios, must be considered.

Uncertainty sources must be sought out. Uncertainty must be assigned to the data. Many input statistical parameters are inferred from the data; uncertainty must be assigned to these. Modelling decisions and parameters must be specified for each algorithm, and uncertainty must be assigned to these too. Measurement and parameter uncertainty transfers through to the uncertainty in the resources or reserves. The sensitive data, parameters, or modelling choices are rarely known ahead of time; they are observed after we have sought them out.

There is uncertainty everywhere: the precise location of the data, its processing and interpretation, our understanding of geological processes, parameters to the geostatistical algorithms, and so on. But after seeking out the critical aspects of uncertainty, our current best assessment must support time-sensitive and business-critical decisions. You have to stop somewhere.

References

Pyrcz, M J and C V Deutsch (2014). *Geostatistical Reservoir Modeling*, 2nd edition. Oxford University Press. 433 p.

In praise of paper and pencil

Nigel Banks

Pencils, ink, chalk, paper, and glue are great map-making and thinking tools. So too are internet mapping sites and GIS software. Choose appropriate tools based on what you need to do.

Krygier and Wood (2011)

This quote from Krygier and Wood's excellent book, which is a must-read for all geoscientists, inspired me to write this article. It reminded me of all the times when I was having a hard time interpreting and felt the need to see my data from a different perspective. One way forward is to shut down the software and do something completely different for a few hours. Another is to print out the key working materials to take advantage of the flexibility that working with paper and pencil gives us.

Paper and pencil are helpful at various stages of many of our geoscience activities. Rather than relying on just one tool, the iteration between paper and software allows us to maximize both our creativity and efficiency. I highlight here the value of this iteration in preparing presentations and reports, in well correlation, and in seismic interpretation.

Presentations and reports

Even Microsoft says that the best way to start preparing a presentation is not to open PowerPoint. They recommend using Word, but paper is much more flexible. The key to preparing a presentation or report is to decide what your story is and then to set that story out in a logical fashion. Jotting a sequence of headlines and sketching out the diagrams, charts, and tables you will need can be done most easily with paper and pencil. Paper also gives us the maximum flexibility to erase, re-write, re-order, and re-number our thoughts until we are happy with our outline. Only then should we turn the computer on.

Additionally, I find this planning is often best done outside the normal working environment, anywhere where your thoughts are free to roam most widely. It might be at home, on a train (where I wrote the first draft of this article), at an airport, or on a plane.

The need for a second phase of 'paperwork' often becomes apparent part way through preparing a large presentation or report when you can no longer see the wood for the trees. It can be a struggle to present your ideas coherently. Try printing out the presentation, and then find a pencil. Slides can often be printed six to a page because it's the flow of the story, not the detail, that is important. A paper copy of a report allows us more readily to see the big picture. Pages can be ordered and pencil notes allow us to re-structure our text and improve the story.

Well correlation and seismic interpretation

Correlating anything more than a very small number of wells requires significant skill in 3D spatial awareness. To assess their geological credibility, well correlations may need to be checked against a variety of maps such as gross isochore or net sand maps. If faults are present we may need to construct fault plane maps. Having these maps in paper form alongside our computer-based correlations makes the iteration from logs to maps and back again much easier than trying to do it all on the workstation.

Correlations are to some extent model-driven and these benefit from being viewed at multiple scales. However large our monitors and however good their resolution they do not provide the flexibility of having paper copies at various scales. The nature of visual perception is such that the patterns we see are scale dependent: by looking at our correlations at different scales we may see a variety of new geological possibilities representing different scales of reservoir architecture. Paper also allows us to test new ideas by quickly colouring different patterns on our correlations, or cutting our paper to re-order wells or re-datum them. These new ideas, if useful, can then be fed back into our workstation interpretations.

The same idea of scale dependency is true for seismic interpretation. The workstation constrains how seismic data can be displayed and we may miss seeing the big picture. Plotting key seismic lines and hanging them on your office wall is a great way of generating new ideas. I remember once visiting my company's head office and looking several times at a regional line hung up in a corridor. I knew it showed something interesting but I could not see exactly what it was. However, on returning to the regional office where I was based I suddenly had a eureka moment. Within a day a new play had been developed and mapped on the workstation, leading to an oil find in a large stratigraphic trap. Without that paper copy it might never have been found.

References

Krygier, J and D Wood (2011). *Making Maps: A Visual Guide to Map Design for GIS.* 2nd edition. Guilford Press. 256 p.

Is working in oil and gas immoral?

Jesper Dramsch

Someone asked this question on *reddit.com*. According to Wikipedia, morality — from the Latin *moralitas* meaning character or proper behaviour — is the differentiation of decisions and actions between those that are 'good' (or right) and those that are 'bad' (or wrong).

There's no doubt the industry has a bad image. It provides fuel for cars, planes, and ships that is burned up and, in the form of carbon dioxide, plays a significant role in climate change and the human effect on global warming. This link alone is pretty bad for the reputation of an industry. If we look further we find accidents such as the *Deepwater Horizon* and ExxonMobil *Valdez* incidents. In addition, the movie *Gasland* fans the fear of fracking. And the global nature of the business is contrary to a lot of people's values and beliefs.

Why would you want to work in oil and gas? One of the first points that come to mind is money. '*Pecunia non olet,*' one might say. With starting salaries in the high five- or six-figure range, you have a major incentive. On top of that, depending on the market situation, the job security isn't too bad if you play things right. Quite a contrast to academia where you likely have a low four or five digit salary and limited-time contracts. And we should not forget about technology — in big companies you have access to the best tools and data.

When you work in oil and gas you can often see how the companies are 'greening up' the business. Now, there are a lot of dimensions to this. The first thing that comes to mind is that they just do it for the image. Being green is good for a corporation's image and it's good for morale. (I don't know one person that deliberately wants to destroy the environment.) Additionally, being green often saves you some cash. If you tell your employees to switch off the computer to save a polar bear from drowning, in the end you will reduce your energy bill and save a lot of money. But there is another big point. People are actually joining oil and gas companies to change things for the better. They want to reduce the environmental impact of that particular company and they have the motivation to change things.

Check your values

- You're against oil and gas. You ride your bike to work and in general you are

Integrity and loyalty to one's values are an important part of morality.

very aware of your carbon footprint. However much an oil company does for sustainability, you will most likely be happier working somewhere else.

- You're against oil and gas, drive your SUV to Walmart, buy global brands, and fly to exotic locations. Please check what you're doing before criticizing someone for working in petroleum.
- You work for oil and gas, you work for the sustainability of your company, and reduce their environmental impact. You will not exactly be green but you will do something about the impact of one of the biggest influences on our environment.
- You work for oil and gas, but do nothing about the environment. If that's in line with your values, it's okay — but don't argue against the impact of oil companies on the environment.
- You just do it for the money. Your values don't come into it. Your employer is giving you an opportunity to change something about its practices but you play along with the status quo. This is problematic in more ways than just working for oil and gas.

My personal opinion

Personally I find it immoral to destroy the environment for profit. Oil and gas are important and valuable resources that should not be burned up like we do at the moment. They're used in the manufacturing of almost everything, and are even essential in a lot of medicines. We need them, but we need to change the way we produce them.

Integrity and loyalty to one's values are an important part of morality. Working in oil and gas can be done in a moral way if you stay true to your values, work towards a more sustainable solution, and reduce your own impact on the environment. You're a small fish in a big pond, but you have a lot of allies and you can effect change on a lot of different levels.

Acknowledgments

This essay first appeared in September 2013 in an extended format as a blog post: *ageo.co/1701Hyq*

It's not about you

Alex Cullum

A joke originally aimed at sedimentologists, but which could realistically refer to many geological disciplines, suggests that all you need for a good argument is two experts together looking at the same rock. Geology has a lot to do with experience — indeed, many claim the best geologists are those who have seen the most rocks. Experience of outcrop, core, and data is obviously vital, and without the ground-truthing this allows, our models would be just elaborate daydreams. Despite emphasis on analogs and real-world examples, either ancient or modern, opinions and belief systems develop. Models appear to give a solid explanation or accurately predict the hypothesis being drilled or tested. Heroes or hero teams, super-theories or paradigm-shifting systems are born. The resulting confidence is liberating, but the egos perhaps are not so inspiring.

Scientific debate occurs between two or more parties discussing points of view based upon evidence. An intelligently presented view can be respected and valued by all parties. But the ego effect erodes respect and others become seen as contemptible fools for daring to question the accepted wisdom. This scenario undermines the whole approach that science depends upon. It is valid — and vital — to question anything at any time. Only by preserving this right will science progress. As soon as there becomes an aggressive or defensive 'them and us' situation, the potential and quality of the science is at stake. Once the aim becomes beating the other team and proving them wrong, the focus has shifted into a dangerous grey area where questionable scientific principles lurk. One tell-tail sign that poor science may be at work is when the source of an idea, theory, or model defines the quality, rather than the value or content, of the work itself.

Time to move on

Addictions, beyond caffeine and chocolate, are another issue slowing scientific progress. We are quick to build models to simplify the complexity of the geological subsurface. But we fall in love with these models too easily and become addicted to their beauty and ingenuity. We're all guilty of thinking our current project is the most important thing, as we dig down into our academic hole, losing sight of the bigger picture — the only thing that might provide a valid comparison of value. As I assert in *Get a helicopter not a hammer*, models are

Science may appear to move forward slowly,
but perhaps it's us who are slowing it down,
by the way we're trying to push it ahead.

best viewed with your helicopter: jump in and fly around, look both up close and from a good height.

Models are just tools for comparison and to be of any use they should not be so precious that when challenged by solid evidence you are unable to accept and adapt your beloved baby or even discard it to build a new one. The mental experiment of using a helicopter can let you dive down into the data to build a scientific case, but then fly up high to get the bigger picture and compare the outcome at scale and within the demands set by the real world. We are typically so blinded by love for our projects and models that we need to get help. Show your ideas, models, and pet theories to others as early as possible. Show and tell even before you are comfortable and ready to do so. Only by failing fast will you be free from the love and addiction to a point where you can move swiftly forward onto the next iteration which might truly be great and worthy of the time you'll invest. Quality assurance isn't something you wait to do until you've burnt lots of time and money. It is an ever-present best friend who you can trust to tell you when you're barking up the wrong tree and it's time to move on.

How many times will we make the same mistake?

As humans we're often good at spotting the issues as they occur or in hindsight, but not so good at making the changes, putting safeguards in place, or communicating the root causes to those around us and avoiding re-occurrence. Multiple-loop learning is even harder to achieve than it sounds. It requires genuinely learning from a project and being able to feed that learning back into the process to improve it next time. Instead, it often takes too many iterations to get things right, or we fail to transfer the experience to those who follow — because we typically are not the ones who will undertake the iterations. For some reason it's more fun to moan to those who are not involved and can't influence the situation than it is to focus our efforts on documenting an issue and talking to those who have the power to improve things for the future.

Science may appear to move forward slowly, but perhaps it's us who are slowing it down, by the way we're trying to push it ahead.

Last day on the North Slope

Gary Prost

Alaska, Sunday 6 August: Rained all night and into the morning. Flew the chopper to Marsh Creek. Walked six miles of the old seismic line with Wayne, reading the magnetometer while he recorded. The magnetic storm on the sun ended two days ago and we are finally getting good readings. The sun is shining with a good breeze blowing... maybe it'll keep the mosquitoes down. We decided not to go with mosquito net headgear and shirts. Without wind we have to rely 100 percent on DEET, a (quasi-toxic?) chemical oil (N,N-Diethyl-*meta*-toluamide) you spread on your skin. This causes what looks like windburn, and melts synthetic fabrics — but it sure keeps the mosquitoes and black flies off. Saw ptarmigan and marmot. Found the tundra uneven for walking: it's best if you can stay up on the tufts of grass rather than the wet spots in between. The tundra is made for folks with one long leg and the other short: one for the grassy tufts, the other for the bog in between. The clouds of mosquitoes — of which there are more than 30 species in Alaska — were on the lee side of hills, mainly around willows. By the time we finish we're in a driving rain with 20-knot winds out of the northwest, temperature 36°F [2°C].

At lunch we went fishing in a rain-swollen, silty river. Fished from gravel bars in the river. Saw a grizzly sow and cub, and I caught five grayling and one char — in Alaska even I can catch a fish. Ed caught eight. We were using spinners with pink spots. It rained the whole time.

Ed and Joe mapped the K–T boundary (about 65 million years old), sampled the Paleocene for source rock, and collected oil from a seep at Manning Point. Today is the first day I really got cold — got wet from inside out, outside in, and bottom up (water over the tops of my mid-calf Sorrel boots). Once back in camp I completely changed clothes, took a hot shower, and collapsed for an hour before cocktails were served in Ed's tent at 6 pm. Dinner tonight (always at 7 pm sharp) is turkey and pork chops.

Two weeks ago we flew into Prudhoe Bay (actually, the town is Dead Horse). Alaska Air lost the bag with my boots; eventually they got it to me and we flew to our camp at the Kavik River on the west edge of the Arctic National Wildlife Refuge. Though we're not allowed to develop, or even explore here, the government is allowing oil companies to do some surveys for a regional study as

The tundra is made for folks with one long leg and the other short: one for the grassy tufts, the other for the bog in between.

long as we don't disturb the wildlife. We worked out of the Kavik camp, flying each day in a helicopter into roadless wilderness where we set down to confirm strikes and dips of bedding mapped on aerial photos, identify rock units, collect samples, and run ground magnetics and gravity surveys.

We heard that a grizzly was killed in the camp as it was being set up in late May: as it wandered near the outhouse someone yelled, 'Bear in camp!' and a shot was fired to scare it off. It didn't scare, sniffed the air as if hungry, and headed straight for a tent where a guy with a bad back was laid up. The camp staff shot the bear just as it got to the tent.

Camp is beginning to thin out. Originally we had ARCO, Conoco, Texaco, Elf, and Placid, plus consultants: about 40 people in all. Now there are only about 15 left, including three camp staff. The sun is out again and is still high in the sky and the camp generators are making their usual racket. Went for a walk after dinner with Ed and Joe along the gravel road that extends 8–10 miles south along the Kavik River.

Finished packing. The little commuter plane picked us up at the gravel strip at 10 pm. We have about 50-mile visibility with low clouds and drizzle. Got to Dead Horse on time and changed planes. The commercial jet left Dead Horse on time. Began reading *Alaskan Bear Tales*, a book I picked up at the airport. A guy sitting near me is wearing a hat that says, 'If you don't think Hell freezes over, you've never been to Prudhoe Bay.'

Learn about seismic

Bruce S Hart

Seismic methods use sound (generally P-waves) to image structural and stratigraphic features. Two-dimensional (2D) seismic data are cross sections through the earth; 3D data are volumes of digital data that, through the marvels of computer graphics, can be visualized in a variety of ways. Most seismic data are collected by the petroleum industry, although other datasets are collected for civil engineering purposes (e.g. aquifer mapping or shallow hazard detection) and academic studies (e.g. deep-crustal imaging).

Here are some reasons why geologists should work with (ideally 3D) seismic data:

You can become a better field geologist. Modern interpretation software allows interpreters to cut through and view 3D seismic data from any direction. For example, faults can be viewed in vertical strike, dip, or oblique orientations, but they can also be viewed on horizontal slices through the data. Even grids of 2D seismic lines cut through structural and stratigraphic features in a variety of orientations. Would you recognize a strike transect through a normal fault as such if you saw one in an outcrop? What does a transect down the axis of an incised valley look like? Working with seismic data will help to train your eye to recognize structural and stratigraphic features in any orientation.

Below is a listric normal fault on uninterpreted (left) and interpreted (right) strike-oriented transects from a 3D seismic volume. Displacement of the hanging wall is directly towards the viewer, but the sense of movement cannot be

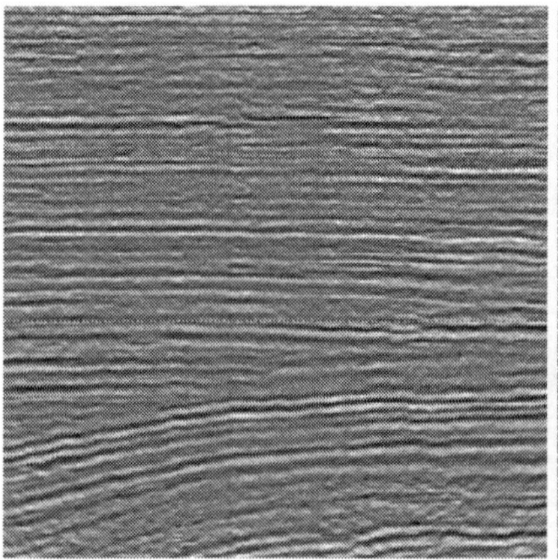

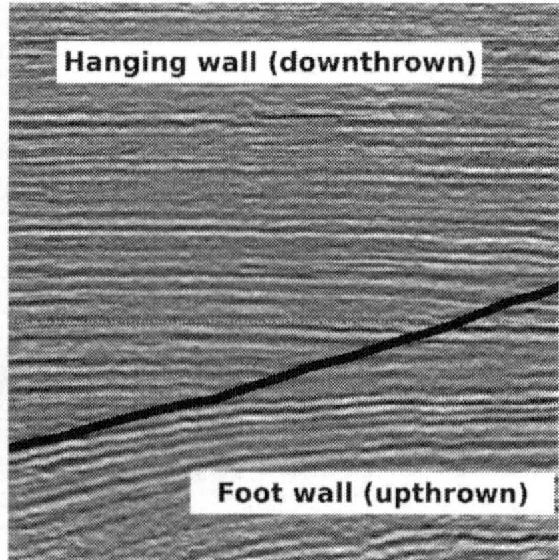

determined from this view. The angular discordance between reflections and reflection terminations could be misinterpreted as an angular unconformity.

You can see more. Seismic data almost always allow you to see more of a structural or stratigraphic system than can be seen in even the most continuous outcrop. There is a lot of geological knowledge to be gained from this big-picture perspective. For example, depositional models for submarine fan systems were revolutionized once high-quality seismic data became available (see *What are turbidity currents?*). The original facies models were mostly developed using outcrop data, but seismic data later revealed that outcrops usually only showed small portions of much larger depositional systems. Similar advances have been made in structural geology, for example helping map linkages of fault systems.

You can map more. You can nearly always make subsurface maps more quickly and more accurately with seismic data than without it. Mapping stratigraphic features (such as channels) or structural features (such as faults) using well control alone is an exercise in connect the dots — each geologist will generate a different product based on his or her biases and abilities. Seismic data, especially 3D seismic data, provide continuous data coverage between wells, virtually eliminating the guesswork involved in mapping solely from well control.

Your career options will be enhanced. It's no secret that most geologists (perhaps excluding academics) work in a boom-and-bust world. Many are the geologists who, over the course of careers spanning a few decades, have changed fields (e.g. petroleum geology to hydrogeology or vice versa) to stay employed. Adding another tool to your kit is never a bad thing.

It can be fun. Most, but admittedly not all, geologists exposed to digital seismic interpretation have come to enjoy working with the software and data.

Despite these advantages, there are of course limitations. For example, in typical data, stratigraphic features need to be at least 10 m thick before they can be resolved; trace-fossil enthusiasts will be disappointed. There can also be geometric distortions to structural and stratigraphic features. To help overcome these and other limitations, and to reduce the ambiguity inherent in working with any single type of data, seismic data need to be integrated with other sources of knowledge — core, wireline logs, outcrop analogs, and so on — during interpretation. That integration effort, in itself, can be an enlightening and enjoyable exercise.

Acknowledgments

Image reproduced with permission from: Hart, B S (2011). An Introduction to Seismic Interpretation. *AAPG Discovery Series* **16**, CD-ROM.

Learn about velocities

Nigel Banks

Petroleum geologists often think that information about seismic velocity is something best left to geophysicists. But there are many reasons for geologists to get involved with velocities. Let's look at two important uses of velocity data: depth conversion and basin analysis.

Velocity models are geological interpretations

The largest volumetric uncertainty for petroleum exploration prospects, and throughout the appraisal and early development of fields, is usually the size and shape of the 'container' — the gross rock volume. This uncertainty might be related to many aspects of seismic acquisition, processing, and interpretation. But seismic depth conversion, traditionally seen as the final step in seismic interpretation, is frequently the source of most of the seismic-derived uncertainty. Even anisotropic pre-stack depth migration does not guarantee the best depth-conversion model.

There is uncertainty (and a need to quantify it) about the relative depths of a horizon across an area: this controls the size and shape of any discovered or potential hydrocarbon accumulation and is required for volumetrics. There is also uncertainty about the depth at a specific location of a certain horizon or horizons: this is required for well prognoses to aid in safe and cost-effective drilling.

Geologists unfamiliar with the principles of depth conversion are often happy to leave this crucial interpretation step to their geophysical colleagues, especially if they see that some mathematical capability may be required. This can be a big mistake. Any variations in velocity in the overburden above your prospect or field are caused by changes in the geology: thus, a depth conversion procedure that isn't founded on or consistent with a sound geological model is unlikely to be optimum, and may be badly wrong.

This is simply an extension of a general principle of subsurface mapping that the contouring of formation tops or rock properties from well data without a geological model is unlikely to give the best map interpretation. The fewer the data points the more important it becomes to have a good geological model. Coincidentally, having a statistically supported geological model is likely to give an even better interpretation — this is why geologists and geophysicists need to combine forces.

Thinking about depth conversion should be a very early step in a seismic interpretation project because we may need to interpret several shallower horizons, possibly solely for depth conversion purposes. Having a clear understanding of the geological history of an area prior to starting detailed interpretation is an essential starting point. Moreover, as more and more seismic data are necessarily processed in depth to handle rapid spatial changes in velocity, understanding how the velocity of the shallow geology varies is crucially needed very early in the seismic processing.

Three things to look for when considering depth-conversion uncertainties are:

- **Low structural relief.** Small velocity variations can cause large changes in the size and shape of low relief closures.
- **Highly variable overburden.** Large water depth variations, laterally varying lithologies (e.g. carbonate buildups, sand-filled channel systems, or salt), and complex geological structure can all contribute to lateral velocity variations.
- **Complex geological history.** Sediment velocities reflect their maximum depth of burial, so the recognition of different amounts of uplift across an area is vital to predicting velocity variations accurately.

So, geologists, talk to your geophysical colleagues about depth conversion and be aware that they may need your help to develop the optimum velocity model.

Velocities can help your geological interpretation too!

Analysing velocities from wells and seismic processing can also provide vital clues to help your geological understanding of an area. In exploration projects we often have limited knowledge of the lithologies present. Seismic stratigraphy and seismic facies analysis can only take us so far. Velocities give us an additional dimension. For example, if the 'pull-up' under a salt dome doesn't seem to be as large as you would expect this might indicate the presence of carbonates rather than clastics in the adjacent rim synclines. Equally, lateral velocity trends beneath an unconformity may help us quantify differing amounts of eroded section and may provide a vital clue to understanding the history of petroleum generation and migration in a basin.

Location, location, location

Matt Hall

A quiz: how many pieces of information do you need to accurately and unambiguously locate a spot on the earth? It depends if we're talking about locations on a globe, in which case we can use latitude and longitude, or locations on a map, in which case we will need coordinates and a projection too. Since maps are flat, we need a transformation from the curved globe into flatland — a projection.

So how many pieces of information do we need? The answer is surprising to many people. Unless you deal with spatial data a lot, you may not realize that latitude and longitude are not enough to locate you on the earth. Likewise for a map, an easting (or *x* coordinate) and northing (*y*) are insufficient, even if you also give the projection, such as the Universal Transverse Mercator zone (20T for Nova Scotia). In each case, the missing information is the datum.

Why do we need a datum? It's similar to the problem of measuring elevation. Where will you measure it from? You can use sea level, but the sea moves up and down in complicated tidal rhythms that vary geographically and temporally. So we concoct synthetic datums like Mean Sea Level, or Mean High Water, or Mean Higher High Water, or… there are 17 to choose from! To simplify things, there are standards like the North American Vertical Datum of 1988, but it's important to recognize that these are human constructs: sea level is simply not static, spatially or temporally.

To give coordinates faithfully, we need a standard grid. Cartesian coordinates plotted on a piece of paper are straightforward: the paper is flat and smooth. But the earth's sphere is not flat or smooth at any scale. So we construct a reference ellipsoid, and then locate that ellipsoid on the earth. Together, these references make a geodetic datum. When we give coordinates, whether it's geographic lat–long or cartographic *x*–*y*, we must also give the datum. Without it, the coordinates are ambiguous.

How ambiguous are they? It depends how much accuracy you need! If you're trying to locate a city, the differences are small — two important datums, NAD27 and NAD83, are different by up to about 80 m for most of North America. But 80 m is a long way when you're shooting seismic or drilling a well.

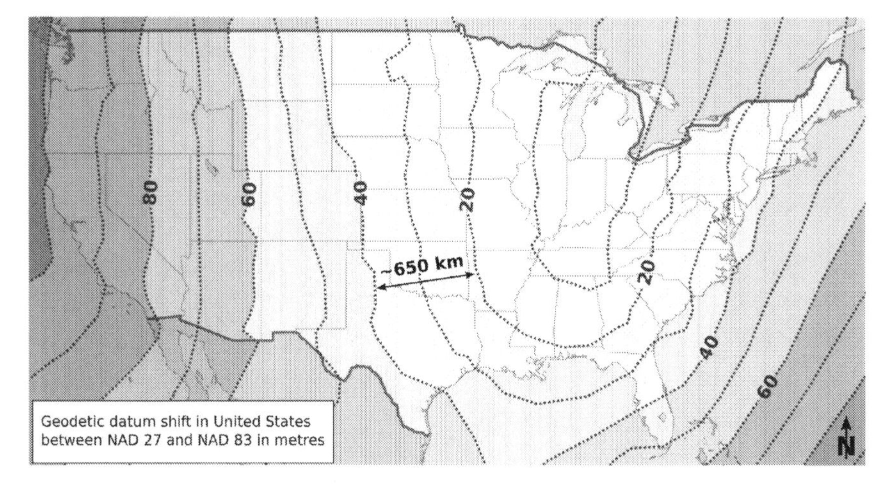

Geodetic datum shift in United States between NAD 27 and NAD 83 in metres

What are these datums then? In North America, especially in the energy business, we need to know three:

- **NAD27 — North American Datum of 1927**, based on the Clarke 1866 ellipsoid and fixed on Meades Ranch, Kansas. This datum is widespread in the oil industry, even today. The complexity and cost of moving to NAD83 is large, and will probably happen v e r y s l o w l y. In case you need it, there's an awesome tool for converting between datums at *ageo.co/17EKX2c.*
- **NAD83 — North American Datum of 1983**, based on the GRS 80 ellipsoid and fixed using a gravity field model. This datum is also commonly seen in modern survey data — a big problem if the rest of your project is NAD27. Since most people don't know the datum is important and therefore don't report it, you may never know the datum for some of your data.
- **WGS84 — World Geodetic System of 1984**, based on the 1996 Earth Gravitational Model. This is the only global datum, and the current standard in most geospatial contexts. The Global Positioning System uses this datum, and coordinates you find in places such as Wikipedia and Google Earth use it. It is very, very close to NAD83, with less than 2 m difference in most of North America; but it gets a little worse every year, thanks to plate tectonics.

To sum up: always ask for the datum if you're being given geospatial data. If you're generating geospatial information, always give the datum. You might not care too much about it today, but once you have spent days trying to unravel the location of a well this will change.

Acknowledgments

A version of this essay first appeared as a blog post in March 2013, *ageo.co/16NLJu3*. The image is licensed CC-BY by Wikimedia Commons user Alexrk2.

Make better maps

Matt Hall

Everybody knows that maps need a scale, a north arrow, and a legend. It's reasonable advice, but there's more to good maps. Here are 11 things you can do to make yours even better.

1. **A scale is obligatory.** That bit of the old advice was good. It's a good idea to show the scale in a couple of different ways — perhaps coordinates (which also serve to show the location) and a scale bar. Make sure you include metres or kilometres — so that non-Americans can understand your map. Including some cultural features (towns, buildings, and so on) helps people intuitively tune in to the scale.
2. **A north arrow is also obligatory.** But the north arrow is not a substitute for showing the regional location and context of your map. Remember that most people assume north is at the top, so if it isn't, make sure you make this especially clear.
3. **Try to avoid needing a legend.** Beware of making your reader look at the legend in order to decode every single map feature. Use standard symbols and colours whenever you can, and label other things directly whenever possible. If colours are available, use them to connect related features to each other and to their annotations or labels. If you are presenting a map digitally, give people a way to query the map layers directly, such as with mouse-overs, without cluttering the map itself.
4. **Annotate the highlights.** The spatial context the map provides will help your reader understand the story in your text. The most salient points of the map are probably obvious to you (see how the azimuth of fracturing changes so abruptly!) — help your reader get excited about them too. But don't go overboard: if everything is highlighted, nothing is highlighted.
5. **Show the database.** It's almost always a good idea to show the data that the map was generated from — wells, seismic lines, outcrops, and so on. This tells the reader which parts of the map are constrained by data, and which are interpolated by algorithms or concepts. You can show this information on a different visual 'plane' by using a semi-transparent grey layer.
6. **Name the projection.** If your map is a projection (that is, flat), then you must give the type of projection and the coordinate system you are using.

Any time you publish or share spatial data, you must give the datum, even for geographic coordinates like latitudes and longitudes. Your data are useless without this information (see *Location, location, location*).

7. **Learn more about colour.** The biggest trap people fall into is using an overly colourful palette for their data, such as the classic rainbow colour scale. Such scales make the map hard to interpret, are hard for colour blind people to use, and leave very few colours for symbols and other annotation. There are plenty of better alternatives — read up on advice about colour, for example on the Rob Simmon's *Elegant Figures* blog (*ageo.co/18hYVY2*) or Matteo Nicolli's *My Carta* blog (*mycarta.wordpress.com*).
8. **Present live data.** Dropping screenshots into PowerPoint misses the point of working on computers. If the situation allows, you'll make a better impression by showing your work live, so that it can be used as the powerful, flexible evidence it is. Your audience can ask questions, you can change scales any time, and you can interrogate your data live. (If the thought of this terrifies you, perhaps your story isn't quite as convincing as you think?)
9. **Appreciate good maps.** Next time you're reading a book or journal, notice which maps you find especially attractive or useful. Conversely, find your least favourite map, and list the things you don't like about it.
10. **Remember your reader.** The point of your map is to inform, and to support your conclusions. Ask yourself — or better yet, ask other people — what they need to know from your map. Don't clutter it with unnecessary details, and make certain the key data are clear and accurate. If the same map must serve two points, consider making the map twice, with different emphasis.
11. **Be a good steward of your data.** If your organization has a process for archiving map data, find out about it and use it. If they don't, be consistent and rigorous, and encourage your colleagues to do the same. Shapefiles are the closest thing we have to a standard map format. When you publish, consider sharing your map data with others by attaching an open license and putting your files somewhere others can reach them — help make subsurface geoscience research more reproducible.

These tips are necessary but insufficient for producing great maps. Don't forget you also need reliable data, sound geological modelling, and a clear idea of the evidence you wish to present. Those are the hard parts!

Making predictive models

Derik Kleibacker

Constructing geological models requires a mixture of deductive and inductive reasoning. There may not be an absolute solution to every level of geological question, but there is an absolute solution to a properly framed geological hypothesis at the resolution the data allows. Answer what you can deterministically and implement probability when you need it.

Deductive thinking dominates the foundation building of a testable hypothesis: *X* equates to *Y*, *Y* equates to *Z*, therefore we predict *X* equates to *Z*.

The pitfalls here lie in the validity of the inputs that are used to fix 'proven' points in your geological model. These deductive assumptions become the trunk of a hypothesis from which branching probabilistic ranges will be leveraged.

Inductive reasoning takes over when significant uncertainty enters the problem: *X* usually equates to *Y*, *Y* usually equates to *Z*, therefore it is likely that *X* equates to *Z*. Even though the input or fixed points in the hypothesis may be sound and the logical argument for the predicted outcome is valid, the result may still not agree with the prediction. An educated range of solutions is required. Multiple working hypotheses fit in this space.

Biases

Bias in geological predictions can come in many forms, and will be the reason the geological model is not useful, fails to identify the critical risk in the system, or fails to capture the range of uncertainty in outcomes.

Deductive failures include misuse of geological principles underpinning the basic assumptions of the predictive model, not recognizing the full range of uncertainty inherent in imperfect geological datasets, and failure to utilize all deterministic geological constraints in the given datasets. The balance lies between recognizing that there are often dead-ends to our thinking that need to be culled, while not pruning the wandering but promising branches of the hypothesis back too far.

Sometimes the interpreter is biased by relying only upon information that is readily available or most easily accessible. For a subsurface geologist, this pitfall could include not being able or willing to interpret seismic reflection data. The

Spend as much time as possible on ideas — building, testing, iterating — and spend only enough time on software as needed to help you create and execute better ideas.

lowest risk predictive geological model can only be achieved by the integration of all available datasets.

There is a tendency to try to confirm rather than disprove a preferred geological model. After creating an elegant, novel solution that fits a preponderance of the data there is a desire to see it proven true, nevermind that nagging piece of contrary information. Geological prediction is non-unique. Don't hide the weakest pieces of your story, highlight them and explain the counter-arguments. Given the many potential outcomes, quite often a geological hypothesis can be harder to prove than to disprove.

Another common bias is the inclination to perceive order where it has not been proven to exist. We seek to find the rules that explain complex spatial and temporal processes that allow us some understanding of the depth of time and earth processes. As geologists we must seek to find these rules, but we must not overreach.

Some ways to get better

- Some will seek subsurface solutions and certainty primarily through application of the latest technology. Spend as much time as possible on ideas — building, testing, iterating — and spend only enough time on software as needed to help you create and execute better ideas.
- Balance the benefits of collaboration with the dilution of groupthink. Curate all available ideas.
- By employing critical examination techniques to facets of your geological interpretation, such as structurally balanced cross sections and chronostratigraphic charts, you will be able to assess your model's feasibility and critical weaknesses.
- No prediction is complete without some measure of its related uncertainty.

Maps are interpretations

Clare Bond

Learning to make geological maps is one of the most fundamental elements of any geology training course. But when you make a map what are you actually doing — creating a document of truth, or a fiction of your imagination? The reality is perhaps a combination of the two. To new geological trainees this may be a shocking admission, but geology as a science is all about interpretation and reasoning (Frodeman 1995). The fact that there isn't a right answer introduces challenges for both the trainee and the tutor. Geology undergraduates are taught and assessed in a world where there are right and wrong answers; changing this perception of education and learning is a process that takes time and experience (Bond at al. 2011).

Think about the process you go through when making a geological map in the field. From the moment you walk, map in hand, into your field area you are collecting information from the landscape. All this conscious and sub-conscious information is combined to build your geological map. There are specific data points based on outcrop which may include quantitative measurements of strike and dip, and qualitative descriptions of the rock structure and properties. Before marking the information gathered in your field notes the interpretation process begins. The colour you choose to mark the outcrop on the map categorizes it into a rock type, an interpretation based on your observations. If well trained you will start making predictions, based on a virtual geological model, from the first outcrop — or from looking at how the rock units intersect the landscape. This initial model is continually tested and refined as the geological map is created.

Critically, although the geological map created is 2D, the virtual geological model in the mapper's head is 3D. This three-dimensionality is what makes the map make sense. Think of the classic children's join-the-dot puzzles. The dots could be joined in any order, but the numbering of the dots allows the child to create a picture. Imagine now your 2D geological field map with no 3D conceptualization or visualization of the geology it represents; you could join the outcrops in any way. But the map contains lots of 3D information, topographic contours, strike and dip measurements of planes, and orientations and plunges of lines. When joining the outcrop 'dots' the geologist uses this information to create a 3D solution represented by the 2D geological boundaries drawn on the map.

Tutors will say, 'Draw your boundaries in the field' — listen to them, they are trying to make it easier for you.

It is much easier to create and visualize this 3D solution in the field, standing in the three-dimensionality of the landscape, than trying to recreate it at your desk. It also allows you to test and build a model in the field. This is why tutors will say, 'Draw your boundaries in the field' — listen to them, they are trying to make it easier for you. Of course the 3D solution created may not be the only possibility, but if it honours the outcrop data and the 3D geometries created make sense, then the solution is a viable geological model and so is the geological map.

References

Bond, C, C Philo, and Z Shipton (2011). When there isn't a right answer: interpretation and reasoning, key skills for 21st century geoscience. *International Journal of Science Education*, **33** (5), 629–652.

Frodeman, R (1995). Geological reasoning: Geology as an interpretive and historical science. *Geological Society of America Bulletin*, **107** (8), 960–968.

Maps are sections

Alan Gibbs

Little of my late 1960s undergraduate coursework was any use in my career as a geologist. The exception was the time we spent in the field mapping outcrops, and a senior staff member impressed on us that 'a map is a section'. Over beer late one night it occurred to me that a geological map is a section not just in 3D space, but it also represents a slice in geological time. That seed idea lay dormant for a few years but then I found a way to build a career and a company — Midland Valley — out of it.

What makes a geologist, and what we are mostly paid for outside of academia, is the ability to tell a story about the geological evolution of an area, and to make maps and sections of the earth's crust at a scale of a few tens of metres to a few kilometres or sometimes tens of kilometres. Mapping and section-drawing skills are at the heart of this, irrespective of whether you are working in engineering, mining, hydrocarbons, or environmental management. In my view, mapping and section building are the core geological skills.

Throughout my career, first as an academic and then in industry, I have been astonished at how poor our skills are at constructing cross sections and maps, relative to other things we do as a profession such as geochemistry or geophysics. Much of what passes for predictive cross sections, while they may honour the data, make no more sense than an Escher staircase. A typical specimen looks good in 2D but makes no sense in 3D let alone in 4D as it has evolved through geological time. If we are to look after our planet more effectively, given our need to exploit it to maintain our societies, understanding the subsurface is more important than ever.

A map really is a section. Any geologist should be able to draw an accurate section across the map that is compatible with it and the 3D object it depicts. The key is to know the stratigraphy so you know what's up and what's down. Then work out the plunge of any structure depicted and the dips of the major structural panels. Don't forget that rule of first-year class — if the surface dips up the valley it will 'V' up the valley. Then the key trick is to place yourself so that you are looking down the plunge and then the map will project onto the section. Try this by slicing up a cardboard tube with a sharp knife. Once you've mastered this concept you should be able to take a map and sketch a cross section across it in a matter of minutes.

If you are looking at a section plane that is not perpendicular to plunge, commonly a strike or oblique section, all the dips will be apparent dips and less than true dip. A lot of geologists get this wrong, especially when drawing faults. If, for example, your section crosses a normal fault perpendicular to plunge then you will see the maximum dip of the fault say 65–70 degrees. If your next section is perpendicular to the first section, the same fault could appear nearly horizontal (see *Learn about seismic)*. All of this uses standard projection and graphic techniques that tend to be relegated to undergraduate mapping classes and then forgotten. Powell (1992) probably provides the most comprehensive guide with numerous exercises ranging from the simple to fiendishly complex.

Using any starting point from a conventional map, it will usually be possible to build a number of seemingly plausible sections. Using a synthetic dataset, Bond et al. (2012) showed that professional geoscientists come up with a wide range of interpretations and only a few arrive at the 'correct' solution. She also showed that those who checked the section by thinking about how the geology developed were three times more likely to be correct than those who did not.

Here are some of the quick checks of geometry and geological sense that I use when drawing sections. This is a process known as 'balancing':

- Do bed lengths add up across your section?
- Can you jigsaw your model back together and understand how the folding might be associated with faulting?
- If you see dip changes across faults, check whether this is due to a shaped fault or if the displacement across the fault is oblique to the section.
- Check that fault displacements vary systematically.
- Check that area or volume changes are compatible with your understanding of the geological history.
- Is there an alternative hypothesis that could help explain the geology?
- Is your model compatible with your understanding of the regional picture? Discrepancies need to be resolved.
- Single sections across strike-slip faults or shear zones won't balance in 2D. On the other hand, make sure you don't just turn everything into a strike-slip deformation zone!

References

Bond, C, R Lunn, Z Shipton, and A Lunn (2012). What makes an expert effective at interpreting seismic images? *Geology* **40** (1), 75–78.

Powell, D (1992). *Interpretation of geological structures through maps: an introductory practical manual*. Longman Scientific and Technical. 176 p.

Maps belong in a GIS

Mark Dahl

Successful exploration programs depend on good decisions being made as soon as possible. Simple enough, but without efficient ways to evaluate and communicate the complex parameters surrounding exploration risk it can be very challenging indeed. Fortunately, there are many software tools available to the industry that make such decisions not only possible but relatively easy. One such tool has been used to solve problems of this type by many other geospatial disciplines for decades — the Geographical Information System or GIS.

A GIS is a computer system capable of assembling, storing, manipulating, and displaying geographically referenced information. It works in much the same way as an Information System one might find at a bank or hotel, only, in addition to referencing an account or file number, information is also referenced to a point on the surface of the earth. This spatial awareness allows the user to perform tasks impossible to achieve with charts and lists.

Although the GIS is a ubiquitous fixture in the energy industry, it is generally poorly understood and rarely applied with effectiveness. Even basic geoprocessing (the sort of functionality that leads to good decision making) tends to be clumsy in the industry-oriented application suite. To say that the GIS has not realized its potential in E&P is a gross understatement. Software developers have not yet successfully closed the rift between the geoscientist and GIS. The reasons for this failure include:

- A significant technology gap between geoscience practitioners and their tools.
- A reluctance to adopt digital workflows and numerical techniques.
- A lack of awareness of how a GIS can be applied to a given problem.

Geologists, who probably stand to benefit from a GIS more that any other geoscientists, are often the most reluctant GIS users. This has to change. Here are eight reasons to start using a GIS:

1. **Capture and storage of data.** Scientific studies — core analysis and description, petrographic interpretation, well tests, and so on — are often not integrated into a prospect evaluation because the geoscientist is unaware of the data's existence, not comfortable with source uncertainty, or just does

not have time to dig through dusty file folders. A GIS is a perfect place to capture and store these data.

2. **Interrogation of data.** A great advantage a GIS enjoys over a regular database is functionality for map-based queries, which give you a means to intuitively interpret and synthesize spatial information while interrogating data. And like any properly organized database, there is effectively no limit to interrogating a spatial database. If you can think it up, it can be done.
3. **Data analysis.** A GIS allows the user to simultaneously display point, vector (lines and polygons), and raster (continuous) data, which are called data overlays. Data overlays are particularly effective for qualitative analysis and illustration. A GIS also provides the user with tools for quantitative analysis, such as surface attribute calculation, map algebra, and geostatistics.
4. **Integration of interpretation.** Interpretation culminates in a series of maps — seismic attributes, horizons and faults, petrophysical properties, stratigraphic gross depositional environments, and so on — describing prospectivity and geological risk. There is no better place to synthesize maps than a GIS.
5. **Validation.** A technical model is based on spatially limited data. An interpreter extrapolates away from control points to predict the unknown. As new data is found or acquired, the model can be validated in a GIS.
6. **Workflow capture and automation.** An interpreter spends a shocking amount of time on repetitive tasks. Thankfully button pushing in a GIS does not need to be one of those tasks. Workflows can be automated in a GIS with easy-to-use programming interfaces. This also ensures that a workflow is executed exactly the same way each time.
7. **Presentation-quality maps and graphics.** The original GIS was developed by a cartographer. Production of beautiful maps has always been a principal objective of GIS. Good ones, such as ESRI's ArcGIS or the open-source QGIS, allow the user exceptional authorship over cartography, including colour palette, line quality, font, and page layout for printing.
8. **Archiving.** Maps and all of their supporting data and interpretation can be neatly organized in a GIS project. A well-organized map project reads like a report, with related material nested in a hierarchy of layer files. Archiving in this manner improves the chances of continuity and integration of technical work between teams in an organization.

There are alternatives to a GIS. However, no other information system — analog or digital — is as comprehensive a toolbox for making better decisions about spatial problems. Your maps belong in a GIS.

Old wells are gold mines

Dan Hodge

There are few frontier areas left to explore. Chances are, the area you're working has had multiple phases of exploration by a number of companies, with a variety of mindsets. It's a challenge to bring something new to an area that has already been picked over. Diligent integration of geology and geophysics is a must, but what has really helped me create opportunities, in a short time frame, is to build as complete a dataset as possible. What I find lacking most often is the integration of old data.

Wise old-timers will tell you the old wells are the best. In fact, new wells are potentially superior, assuming all logging runs went smoothly and the special analysis comes back in time for your next decision or well location. But old well files provide instant gratification, months if not years of work in one data package. They can de-risk a play, make (or break) a prospect, save you money, and can even enhance your reputation. They are worth more than their hardcopy weight in gold and it surprises, in fact inspires, me when old data is shrugged off as irrelevant.

Not only are old wells cheap, there is a chance you may find the data you were unsuccessful in persuading your manager, drillers, or the borehole to provide for you in your exploration campaign. New core can be especially hard to come by.

One reason old wells get overlooked is the wellfile is incomplete. Another reason is the well is completely missing and without further investigation you would not even know it existed. The older the well, the greater the potential for missing data. Wells drilled during exploration booms sometimes suffer the same affliction as filing becomes less diligent.

I have never regretted the extra effort put into bolstering my database with old, lost, or apparently missing data. A better database than your peers or predecessors will give you a good advantage. Here are some tips and advice to make the most of this inexpensive resource.

Seek the data. Persevere: missing data is missing because the last person gave up. Data has turned up in random government departments, misplaced in other wellfiles, a box of unscanned files… the list is endless. Keep a missing data

I have never regretted the extra effort put into bolstering my database with old, lost, or apparently missing data.

list and revisit it from time to time as your understanding of data repositories evolves. For the most part I've found the data is not lost — it's hiding.

Missing wells. I've always found at least one 'surprise well' in each block I have worked. I recommend you surprise yourself early before any exploration campaign. An effective method I use to check for missing wells is to overlay maps of different sources and vintage; this also is an easy way to check well locations. Another good way to reveal missing wells is to plough through both the geology and drilling and completions section of available well reports. Often references are made to past drilling campaigns. These reports usually reference the data collected — a quick check to establish if you have all the logs or special analyses available.

Location, location, location. Check locations no matter which country you're in, what company drilled the well, or who has made the map. I've found high-profile wells 1 km from the actual location. Google Earth is brilliant for this kind of check. I've found drilling pads (squares approximately 50 × 50 m) for wells as old as 1955 in dense jungle. I recommend you have locations checked in the field. Another tip: X marks the spot. Diligent seismic operators will scout old well locations to tie with the seismic, so if you have conflicting locations for a well check to see which one is intersected by two or more lines.

Integrate and scale. Have image logs digitized or depth registered — viewing the logs in the same vertical scale is a must, and comparisons of logs is far easier when they are in the same lateral scale. Resistivity curves (usually the most consistent in old logs) are usually presented in linear scale. If I don't have time to digitize, I will load the image into the cross section and present the digital resistivity curves in a linear scale. Linear scale resistivity is actually quite a useful correlation log in formations with less dramatic resistivity fluctuations.

Economic resolution. Times have changed since the well was drilled, thinner beds can pay out and so can tighter formations. In light of this, review the old wells. Do this at least every 10 years, or every time there is a paradigm shift in the industry.

Pemberton's laws of stratigraphy, part 1

S George Pemberton

As a professor, I am often asked to give advice to students who are going into the petroleum business so I came up with 31 laws that I believe all geologists need to keep in mind. I hope they make young geologists think.

Here are the first 16 laws:

1. I generally start all my classes with the Chinese symbol for chaos and the saying: *Before there is understanding there must be chaos.* I tell students to fully embrace this concept because they must sift through a lot of material that at first they find confusing. If they are willing to embrace this chaotic mess, it slowly starts to make sense and they will be able to master more abstract aspects of the material. It takes hard work, and the only way you can accomplish this is if you sit down and work through the material until it isn't overwhelming any longer. Because of this, I end my classes with the same Chinese symbol but now state *Chaos: Where brilliant dreams are born.*

2. Geology is simple: You have a stratigraphic framework, the mineralogy and paleontology of the stratigraphic framework, and the structural deformation of the stratigraphic framework. Everything must be related to the stratigraphic framework or it has limited value.
3. You can't model nature. There are just too many variables.
4. You can make a model based on your observations. Never let a model dictate your observations.
5. There are no unique indicators in geology. As soon as you adopt the 'this means that' philosophy you have lost the war.
6. The best geologists are the ones who have seen the most rocks. In geology

there is no substitute for experience, so seek it out. When you first start your career it's up to you to find your own mentors, then keep your mouth shut and your mind open.

7. Geology is not an exact science — it is an interpretive science. Do not be afraid to make an interpretation. Remember though that interpretations evolve and change as more data comes to the table; do not be afraid to adapt to those changes.
8. Just because it's published does not mean that it's right. Pretty diagrams do not equate to good science.
9. The only facies models that have ever worked are turbidites and the point bar because they can be interpreted hydrodynamically.
10. Although the rock record is static it is representative of dynamic processes. It is our job to unravel those processes. We see a mudrock as a static object and give it one interpretation, but it may represent a dynamic history. The mudrock may have been originally deposited in a salt marsh where it was buried and dewatered then erosionally exhumed in the beach where it is modified, buried, and partially lithified. It may then again be erosionally exhumed as relict sediment offshore. We need to be able to interpret this complex history.
11. The rock record is more gap than record.
12. Stratigraphy is synergistic and it incorporates all aspects of geology. This is especially true when doing subsurface geology. Petroleum geologists must be eclectic and be able to understand and integrate a wide variety of datasets.
13. Every day in geological time is unique unto itself because the position of the continents is slightly different and will never be exactly the same again.
14. The rare event is more commonly preserved; the rock record is a series of catastrophes. This is especially true when dealing with clastic systems.
15. The deliberate search for the subtle stratigraphic trap requires geological expertise. At exploration scale, many large stratigraphic reservoirs can be overlooked on seismic and the geologist must turn to core for reliable interpretations.
16. A surface is not a surface unless it can be mapped.

This list continues in *Pemberton's laws of stratigraphy, part 2.*

Pemberton's laws of stratigraphy, part 2

S George Pemberton

Here are 15 pieces of advice — part of my 31 laws (see *Pemberton's laws of stratigraphy, part 1*) — for students entering the petroleum business, including perhaps the most important of them all.

17. Exploration is finding the anomalies and figuring them out. There are two types of geologist. One sees something odd and shrugs their shoulders and continues on. I want them to work for my competition. The other sees the same thing and asks why and tries to figure it out. I want that geologist working for my company.
18. No geophysical log actually measures grain size.
19. The reservoir is the rock not a squiggly line on paper. If you understand the rock you will understand the reservoir. I take great inspiration from a phrase commonly used by the late Gerry Friedman: *saxa loquuntur* or 'rocks speak'. It is our job to learn their language and listen to what they are telling us.
20. Calibrate geophysical data to core, then use it to interpret wells with no core.
21. A seismic section is not a cross section of the rocks.
22. You need acoustic impedance to get a signal.
23. One man's signal is another man's noise.
24. The best interpreters are the ones with the most vivid imagination. This pertains to both geology and geophysics. Look to people like Robert Weimer and Henry Posamentier for inspiration.
25. Contouring is the greatest skill you can develop. Our business is dictated by maps — know how to draw them by hand.
26. A computer map is a mathematical expression, not a map. Most contouring software does not incorporate geological principles; your geological bias is what you get paid for.
27. All companies have the same toys — software, hardware, etc. What separates companies is the people who manipulate those toys. Arnold Bouma summed this up very succinctly when he wrote, 'There are still discoveries to be made, but it won't be the computer that tells us what it all means. For that, we always have to go back to the rock to find out what we can do with it and what it means. And for that, the geologist who can explore and observe

Don't lose sight of the fact that you are a professional that must continue to learn and develop your craft.

and think is still the most important thing.'

28. Not everything is allogenic. All sharp-based sandstones are not forced regressions, and all channels are not incised valleys. Many systems contain autogenic elements that rely on *in loco* changes in sediment supply, local tectonic events, and so on, to initiate local changes in relative sea-level.
29. There is no such thing as a finished map — it should be in a fluid state of constant revision.
30. Beware the geologist with the same interpretation for everything. This means that they are either pushing a particular model or that they have only worked in one type of system and their bias is pushing their interpretation. You must be open to following the direction that the rocks are taking you.
31. Perhaps the most important law is this: do not contract petroleum disease. This is when you get that high-paying job with all the perks and toys and forget to do your homework. Don't lose sight of the fact that you are a professional that must continue to learn and develop your craft. If not, when the next downturn comes (and believe me there will be a next downturn) you will be the first one out the door.

Plumes do that

Erik Lundin

The plate tectonic paradigm profoundly changed the way geologists viewed and understood our planet. Elegant and powerful as the paradigm was, geologists still had a need to explain volcanism located away from plate boundaries, i.e. volcanism that could not readily be related to processes to do with continental break-up or subduction. Thus the term 'hotspot' was coined by Tuzo Wilson in the early 1960s. Classic examples of such volcanism are the islands of Hawaii, situated within an oceanic plate, and Yellowstone, situated within a continental plate.

Given that hotspots could not easily be related to plate tectonics, another explanation was sought. In the early 1970s Jason Morgan provided the concept of mantle plumes — roughly 3000 km-tall columns of abnormally hot mantle rising from the core–mantle boundary and eventually inducing hotspot magmatism on the overriding plate. Conceptually this is a bit like moving your hand over a burning candle. This concept was not plucked from thin air, since it was well understood that convective turnaround of the earth's mantle was needed to explain the cooling of the earth. The failure to appreciate heat loss via convection is, for example, blamed for Lord Kelvin's underestimate of the age of the earth. So given that convection was accepted, Morgan's plume concept was well received. The number of such plume features was originally estimated to be 20 or so. These rising columns of molten rock were also suggested to be stationary with respect to the earth's core, and so the concept was quickly embraced by researchers reconstructing plate motions, since it provided an absolute reference frame.

The plume concept became so popular that before long there were about 5000 proposed hotspots and associated plumes! Hotspots were no longer restricted to plate interiors and the term was also applied to plate-boundary volcanism. More remarkable still, is the flexibility granted the plume concept.

I recall a geological conference where I met an eminent professor, a world authority on plumes, and asked about his recently proposed plume model. This particular plume was first expressed as a 2000 km-long sub-vertical mantle sheet, inducing an equally long magmatic province. Some 5–10 Ma later, the sub-vertical mantle sheet swung around by 90 degrees, produced another 2000 km-long magmatic province at right angles to the first, and while doing so broke the plate. Following this neat acrobatic trick the plume collapsed into a Morgan-type cylindrical-

A concept that is granted the freedom of perpetual ad hoc *amendments has the ability to explain anything, and is hence attractive to some people.*

shaped plume. I asked if such unlikely acrobatics were not artificially tailored to match surface observations, and received the firm reply, 'Plumes do that.'

Such malleable concepts are useful. We should have more of them in geoscience. Fortunately for us geoscientists, the plume concept has been expanded and can now explain far more than the original idea could. Take the case of Iceland, a commonly cited super plume, supposedly rooted at the core–mantle boundary and by chance intersecting the mid-Atlantic plate boundary. Unlike Hawaii, Iceland does not have a hotspot track, but that is OK because one can calculate the plume's paleo-position since it is fixed to the earth's core. And lo and behold, the Iceland plume can be followed on quite a journey through time and space. Some 250 Ma ago, this plume was apparently responsible for the Siberian traps, after which it took a 120 Ma break before emerging in the Alpha Ridge area of the Arctic Ocean. And guess what? After another 60 Ma pause in the magmatic activity the plates had moved such that the plume again caused basalt extrusion, this time in the Disco Island area of West Greenland. Somehow the plume shortly thereafter sent off a tentacle to the incipient northeast Atlantic, or alternatively the plate moved quickly over the plume. In any event, the plume emerged from beneath Greenland and induced northeast Atlantic break-up, in a magma-rich manner. This magma-rich break-up is itself a true sign of plumes, for what else could cause it? Surely not the plate tectonic break-up process itself? That would just be too easy. The idea that there could be a genetic relationship between rate of plate separation and the amount of adiabatic melting is actually quite appealing to some of us. But no, let's introduce a plume instead. After all, plumes being fixed to the earth's core and remaining so for hundreds of millions of years within a simultaneously convecting mantle is a sign of a robust concept, right? Plumes do that.

More recently it has been suggested that plumes are not strictly fixed with respect to the core, but sway in the mantle wind. That's OK. One has to be a bit flexible.

A concept that is granted the freedom of perpetual ad hoc amendments has the ability to explain anything, and is hence attractive to some people. But such a concept can neither be falsified nor used predictively. In the long run it may be wiser to ask yourself 'Is there an alternative explanation?' rather than simply shrugging, 'Plumes do that.'

Precision is not accuracy, interpretation is not truth

Gary Prost

Despite having been taught the difference between precision and accuracy, geologists often confuse the two. To use an example, we can calculate reserves to three or five or even ten decimal places. However, when a geologist works with a field that has a billion barrels in-place, this amount of precision does not mean the data are correct (accurate). We are putting too fine a point on it.

Before computers there would be a couple of wells in an area and calculating reserves was done with estimates. Inputs included average pay thickness, average porosity, average saturation, and a recovery factor calculated for an interval of interest. We would multiply them to come up with an estimate that might be off by as much as 25 percent (or more). Since this was done by hand we recognized the assumptions involved, and realized the lack of precision and accuracy. We knew that the answer approached the 'truth' as the number of wells in the prospect increased. Now that we use gridded contour maps or build geomodels, and because the computer can give us an answer to as many decimal places as we like, we tend to believe the output as unvarnished truth. Even if we recognize that this 'most likely' case is one of many possible outcomes, we then provide the maps and numbers to management without explicitly stating this.

The correctness of our calculations depends on the accuracy of many inputs, including estimates of porosity, thickness, saturation, and recovery factor. Each of these has an error bar that represents the assumptions that go into correcting the raw logs and interpreting those logs. Each log has a vertical resolution that depends on the specific instrument configuration and design. Corrected logs such as gamma ray, resistivity, and density curves are used to create interpreted logs such as shale volume, porosity, saturation, and perhaps a pay flag that provides net pay thickness. These interpretations make several assumptions such as being 'on depth' and no instrument errors. As well, the log values are averaged when they are upscaled from the 15 cm sample interval of the logging tool to the 1 or 2 metre (or more) cell thickness in a geomodel.

Logs are averaged over the entire pay interval when we make a gridded contour map. When we build a geomodel or make a contour map we assume the values at the wells are 'true' rather than close approximations or averages. We hold the value at the well constant while using some statistical process (e.g. sequential

The precision being used must not exceed the accuracy of the data.

Gaussian simulation) or facies distribution concept (e.g. fluvio-estuarine system) to extrapolate values away from the wells. Sometimes we impose trends on the data when we use a mental model (a preconceived bias). Using variograms in a geomodel, for example, allows us to dictate how far away from a well and in what direction porosity or saturation values can extend their influence. These processes have built-in assumptions about the distribution of rock properties away from wells. The most egregious assumption is that a well or two, each of which samples maybe a 10 cm diameter section of rock, and the logging tools we use, which sense maybe a metre or two into the formation, are representative of an area that is tens of square kilometres or more.

So how does this affect the geologist? You report that your prospect contains 411.6 million barrels of recoverable oil. You know this because your geomodel told you. The following year you drill a new well, or change your variogram distance, or change your thickness cutoff, or decide that effective porosity is a better measure than total porosity in your area, or modify your recovery factor from 40 percent to 35 percent. Suddenly your reserves are 340.4 million barrels recoverable. You just lost 71 million barrels, equivalent to a good-sized field. You have to report this loss to your boss. Worse, she has to report it to her boss. You can bet someone will want to know what happened.

Without acknowledging it explicitly, your boss took your reserve number as 'the truth.' You neglected to put a range of possible outcomes around your first number that took into account all of the estimates and assumptions that went into it. Because of the precision available from your computer model you inadvertently communicated to your boss, or your boss assumed, that the answer must be essentially true. In fact there could still be a 25 percent error in the actual recoverable estimate. Since your second number was about 17 percent smaller than the original, it suggests that both are within a valid range of probable outcomes. The variation is within the background noise of the accuracy of the estimate.

This realization applies not just to estimates of reserves but to every aspect of our work. The precision being used must not exceed the accuracy of the data. We must recognize and communicate that the interpreted answer is merely our humble attempt to approach the truth.

Presenting… your career

Tony Doré

Whether you're in geoscience or accountancy, you're living in a PowerPoint world. This one piece of software dominates the way we approach presentations. For those geoscientists who used to travel to meetings burdened by profiles, maps, and transparencies, the benefits are obvious. A tiny flash drive does the trick now, or we can just pull the stuff off the net. Then there's the flexibility — we can reorder, combine, import, animate, and alter with minimum effort. We should be in presentation Nirvana.

Unfortunately, as with most things digital, the PowerPoint world brings its own pitfalls. There are probably more bad presentations around than ever before, because it's just too easy. Throw a few slides together from diverse sources and you've got a scientific talk. That's why many of our geological conferences are over-subscribed with substandard papers. The tool can be misused, and the casualty is good scientific communication. However, I'm not writing this to bewail the standard of geological presentations. There's a more pressing matter — your career.

Your contact with the power brokers is probably limited. A presentation is a rare occurrence when you, personally, are being showcased. Trust me on this — a single presentation can leave a lasting impression and be pivotal in your career, whether academic or industrial. I've been on both sides of the leadership fence and I know it's true. I've also observed that getting a few big things right is paramount; the rest is purely cosmetic.

Top of my list is this: enthusiasm conquers almost everything, including lack of formal technique. There are courses that claim to iron out wrinkles in your communication skills, but be careful you don't lose what comes naturally. Most of these courses will put you on video so you can see your idiosyncrasies laid bare. You could end up more self-conscious than before. What really counts is engagement with your subject, looking your audience in the eye, and making them part of your passion. If that involves waving your arms about, so be it. I've seen people with severe tics and stammers give storming presentations.

Next: when you rehearse, concentrate hardest on what connects the current slide to the next one. You already know the content of each slide pretty well. But just describing them in turn is like one of those embarrassing after-dinner speeches where the speaker keeps saying 'And another thing….' The links turn a series of disjointed slides into a story, and create a flow that will both engage and impress your audience.

At the same time, think about who your audience is. Are they geoscientists or civilians? Are they management, your peers, or some combination? Then pitch your talk accordingly. 'Polyphase deformation in retreating extensional subduction systems' will impress your colleagues, but will send the average executive into a reverie on last night's big game.

As a geologist, treat raw GIS outputs and 3D workstation screen dumps with caution. Wonderful as these tools are at assembling and representing geological information, without editing they don't usually make very good slides. Remember, your public has only seconds to understand a picture. Simplify maps: you are trying to communicate an idea, not impress people with the complexity of your data.

Carefully tailor your presentation to the allotted time. The old rule about a minute per slide isn't bad. Most critically, try to make sure you have enough time to say what you want to say. We all underestimate the time we need. You've heard it before: 'I can do this in 15 minutes, easy!' And you know the usual outcome. If you're on a strict time schedule, and you have any say at all, try to keep the discussion for the end. It's very easy for a presentation to be derailed by overenthusiastic debate so you never get to make your main point.

Don't revisit and rephrase points, except perhaps in your conclusions. I inwardly groan when someone, having explained a point perfectly well, then says 'So what I mean is…' It sounds like you're having trouble grasping the point yourself, and of course it creates time stress.

Finally, use bullets as cues, not scripts. The bulleted list is a staple PowerPoint tool, but it's critical not to overload it with text. Reading out a wordy point verbatim sounds unprofessional and, because your audience can actually read, they are probably doing so instead of listening to you!

Remember, you have every reason to be confident. You have a tremendous advantage over your audience: They're coming to this cold, whereas you've prepared hard, got a good storyline, and know it inside out. For 20 minutes, you are the world expert. So unleash your enthusiasm and go knock 'em dead!

Rocks don't lie

Tom Moslow

This essay is written for geoscience students who aspire to a successful career in the oil and gas industry. I present a set of guidelines which I venture to call rules. They are based on my 35 years of experience as a professional geologist working in industry, academia, and government. These rules have as much to do with human behaviour and personality traits as they do with geological intuition. They comprise a set of attributes inherent to success in almost any endeavour, but perhaps especially a career as an oil and gas industry geoscientist.

The 10 rules do not constitute any portion of a geoscience curriculum. They are not lessons that would be taught or learned in university. Perhaps they should be; perhaps one day they will be. Until then, use these rules to your benefit.

1. Rocks don't lie. The most sophisticated and detailed analytical, petrophysical, geocellular, statistical reservoir model constructed with the most advanced software on the most advanced computer is not worth a hill of beans if it's not rooted in fundamental empirical observations of the sedimentary rocks that comprise that reservoir.
2. Be passionate about your work. Be passionate about your prospect. Display genuine enthusiasm in all that you do…
3. …but don't fall in the love with your prospect as it can, and sometimes will, break your heart.
4. Be a team player. Nobody wants to work with an egocentric fool. Expect to encounter difficult, if not impossible, personalities. Learn to deal with them but never compromise your own values.
5. Never force your observations to conform to a preconceived notion. Trust your intuition and trust your gut. Challenge conventional wisdom.
6. There can be nothing more painful than the truth at the end of a drill bit. Expect to make mistakes; learn from them and move on. If you fear being proven wrong, you don't belong in this industry.
7. Lateral variability within sedimentary strata is the rule, not the exception. Expect it, understand it, and you will learn to predict it.
8. As a corollary to this, the search for homogeneity in the subsurface leads nowhere, as it simply does not exist. At some level and at some scale, every

Be passionate about your work.

Be passionate about your prospect.

Display genuine enthusiasm in all that you do.

play, even every 'resource play', has its degree of heterogeneity.

9. The best technical prospect may never get drilled — for reasons that have absolutely nothing to do with geology. Ultimately, decisions are business based. Let it go; learn to move on to the next great opportunity.
10. The bottom line is the bottom line. First and foremost, it's a business, and success is all about making money.

While these rules may not guarantee success, they will assuredly inhibit failure. Maybe one day you can pay me a royalty!

Scaling the outcrop

Allard Martinius

Feeling gravity taking over and pulling one down while struggling up the debris aprons formed by pieces of heterolithic Neill Klinter Group (exposed along the cliffs of Jameson Land, East Greenland) is not an encouraging feeling. The cliffs have to be reached though, because the succession is packed with relevant information for analogous hydrocarbon-bearing formations, such as those on the mid-Norwegian continental shelf. Three steps up and two down is the rule; the loose scree material lying at the angle of repose forces you to be patient and persistent. With all safety precautions taken, and most accessibility issues overcome, a tremendously rewarding day lies ahead.

At the end of the day, yet another section has been completed, and numerous samples have been taken and stored. The results — such as qualitative and quantitative data, panoramic photographs, insights, a proper feeling of geological scale — will be highly relevant. They will allow for further reconstruction of an ancient, continuously changing depositional system and its complex geomorphological and stratigraphic architecture. The laboratory of a sedimentologist (or, in fact, any geologist) is an outcrop, all outcrops. Only outcrops can provide the quality and quantity of information and insight with which to better understand subsurface analogs. Just as a biologist or a physicist needs a laboratory to carry out his or her experiments, so a sedimentologist needs outcrops. And yes, the adage is true: the more outcrops you have seen and studied, the more value you add to your work.

Why is it that hydrocarbon-bearing successions in the subsurface at several kilometres depth can never be sufficiently understood without the aid of quantitative and qualitative outcrop data? This data is used to build conceptual models, build three-dimensional, geostatistically based computer models, populate software with realistic information and numbers, test assumptions, and so much more. Although the instruments used to collect information in the subsurface are very advanced, they represent indirect information only. They provide very high-density but spatially discrete datasets in one, two, or three dimensions but they have a large range in resolution and can simply not compete with continuous cliff sections that may stretch for up to tens of kilometres in length and cover hundreds of metres of stratigraphy. All hydrocarbon com-

Relying solely on workstations may lead us to take two steps up but three steps down.

panies supplement their subsurface evaluation with outcrop analog studies because they realize that outcrop information is an essential ingredient for better reservoir description and for finding the most appropriate drainage strategies.

But maybe more than anything else, outcrop studies allow for the assessment of natural variability and uncertainty related to any type of data. Getting it right in one place does not mean you will get it right in another. Underestimating natural variability leads to errors in your work. As long as you are aware of this you will want to see more in order to learn and improve. Therefore, it is essential that you study not just one example of a particular depositional system but several more as well. Although classifications can be made and general rules extracted, natural variability is always larger and we need to understand it well.

The necessary skills for outcrop analysis and data acquisition have to be consistently taught (and re-taught) at our universities so that outcrop analog studies can continue to provide crucial information to the hydrocarbon industry. The industry is still using outcrop analog data widely and exhaustively but cannot provide all the training to acquire the basic acquisition skills. The geoscience community needs to preserve the right balance between much-needed modern computer workstation skills and the equally valuable but more traditional field geology skills. Remember the Neill Klinter slopes — relying solely on workstations may lead us to take two steps up but three steps down.

Strata are not flat

Brian Romans

Students in introductory geoscience courses typically first consider stratigraphy as a pile of horizontal layers or beds. Fundamental tenets of geology such as the principle of superposition and the principle of original horizontality help stamp an image of so-called layer-cake stratigraphy into the minds of incipient geoscientists. Moreover, introductory structural geology classes show diagrams with layers of uniform thickness and character to unambiguously illustrate types of folds and faults.

But this is wrong. It is more appropriate to think of sedimentary layers as sedimentary bodies. No sedimentary layer extends out uniformly across the planet, they eventually pinch out, get cut out, or maybe transition into a different type of sedimentary rock. Thus, these sedimentary bodies have a shape to them, they are not rectangles. Nearly all sedimentary bodies, when viewed in a 2D cross section, have wedge-shaped edges. Bailey (1998) called these bodies 'lenticles' after their lenticular shape. The geometry might not be visible unless you impose a great deal of vertical exaggeration (as stratigraphers often do with seismic reflection profiles) but it is there nonetheless.

The layer-cake view of stratigraphy does not consider:

1. complex patterns of deposition and erosion, or
2. complex patterns of preservation. A snapshot of a depositional system (i.e. some place where there is net accumulation of sediment) looks more like the Landsat image of the Lena Delta on the facing page.

In most cases, however, we do not see a fully and perfectly preserved depositional system in the rock record. Processes related to the preservation of a body of sediment or rock results in an amalgam of numerous elements of the system over a long period of time. These processes can, in some cases, 'smooth out' complexities but they also introduce a few of their own.

Considering stratigraphy as a 3D configuration — or architecture — of sedimentary bodies is important for subsurface characterization. In some cases you might have sufficiently high-resolution seismic reflection data to explicitly map such geometries. In other cases you will be asked to interpret (predict!) them.

No sedimentary layer extends out uniformly across the planet, they eventually pinch out, get cut out, or maybe transition into a different type of sedimentary rock.

References

Bailey, R (1998). Review: Stratigraphy, Meta-Stratigraphy, and Chaos. *Terra Nova* **10**, 222-230. DOI: 10.1046/j.1365-3121.1998.00192.x

Acknowledgments

This essay is modified from a version that first appeared as a blog post: *ageo.co/14Tlbas*

Stratigraphic surfaces are complicated

Zoltán Sylvester

If you have seen the geological spectacle called the Grand Canyon of the Colorado River, your first impression might have been that the stratigraphic record is not that different from a layer cake. Most bedding surfaces visible in the canyon walls are roughly parallel to each other and many sedimentary layers seem to go on forever. However, this picture breaks down if you look more carefully. At a large scale, layers that seem to have a constant thickness turn out to be consistently thinning in one direction and/or they are truncated by large erosional surfaces; at a small scale, as you look closer at some of the rock faces, interesting patterns with inclined bedding and smaller-scale erosional boundaries emerge. The geometry and origin of stratigraphic surfaces and their relationship to ancient land- and seascapes may seem to be simple geometric problems. In fact they are far from simple.

To better understand why, it is helpful to view stratigraphy as the result of cumulative changes in the surface of the earth. The change can occur either through addition (deposition) or subtraction (erosion). While it may be easier to think of the surface — or the landscape — as undergoing either deposition or erosion everywhere at any given time, there is no reason why parts of it can't be eroding while deposition is taking place not too far away. Yet conventional stratigraphic thinking is strongly influenced by the idea of a clear separation of erosion and deposition in space and time.

There are certainly cases when and where this view is valid; classic angular unconformities come to mind. When James Hutton saw the unconformity at Siccar Point, where only slightly tilted 345 million year old layers of the Old Red Sandstone are sitting on top of near-vertical beds of 425 million year old Silurian greywackes, he realized that such structures could not have formed in only six thousand years, widely believed to be the age of the earth in the late eighteenth century. He reasoned that:

- The sediment in the older formation was deposited in horizontal layers.
- It got buried, compacted, and became hard rock.
- It was tilted to an almost vertical position and lifted above sea level.
- It was then eroded by subaerial erosion.

The Wheeler diagram ultimately became the spark that ignited a revolution in sedimentary geology.

- It was buried again by much younger sediment that was itself later cemented and tilted by tectonic forces.

Although the eroded material that corresponds to the erosional surface must have been deposited somewhere, we know that this place was far away from the location of the erosion.

For two centuries after Hutton's discovery of unconformities, stratigraphy essentially remained the study of quasi-horizontally deposited sedimentary layers; the study of deposits — and the fossils they contained — rather than surfaces, with little interest in the nature of erosional boundaries. However, in the 1950s Harry Wheeler, professor of geology at the University of Washington, started to explore ideas about where the eroded material went and what happened to stratigraphic discontinuities if one was able to follow them laterally over long distances. Wheeler realized that:

1. The time gap represented by an unconformity must include both the duration of deposition of the eroded sediments (he called this the 'degradation vacuity') and a time of erosion and non-deposition (hiatus); and
2. This time gap must decrease if we follow it toward the area where the eroded material is deposited.

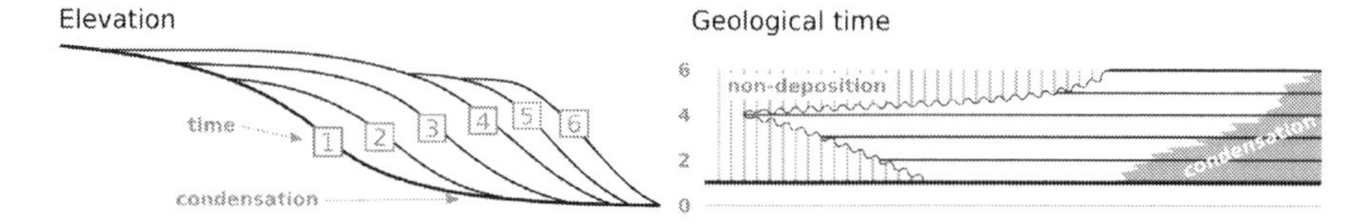

These concepts are best illustrated on a cross-section-like diagram that has time on its vertical axis, instead of elevation or depth. Nobody before Wheeler had looked at stratigraphy using such an 'area–time section', as he called it, and the Wheeler (or chronostratigraphic) diagram ultimately became the spark that ignited a revolution in sedimentary geology.

This essay continues in *Stratigraphic surfaces are* really *complicated.*

Stratigraphic surfaces are *really* complicated

Zoltán Sylvester

Professor Harry Wheeler was one of the architects of modern stratigraphy. His ideas seemed abstract and difficult to many of his contemporaries in the 1950s, and there wasn't a whole lot of data to support them. While his ideas initiated a revolution in sedimentary geology, it was a few more years before the increased availability and quality of seismic reflection profiles offered unprecedented insights into the large-scale architecture of sedimentary rocks.

In the 1970s, using many seismic cross sections from continental margins, Peter Vail and his colleagues at Exxon's research centre in Houston identified a whole range of large-scale stratigraphic patterns and interpreted them as the effects of sea-level change and subsidence on sedimentation and erosion. By doing so, they created sequence stratigraphy, an entirely new way of looking at sedimentary rocks. Erosional surfaces became important in sequence stratigraphic analysis, as stratigraphers realized that deposits of tectonically quiet continental margins have numerous stratigraphic discontinuities that are more subtle than major angular unconformities, yet can be confidently traced across great distances. These sequence boundaries divide the stratigraphic record into sequences. Deposits below and above the boundary have little to do with each other. This is in contrast with a relatively continuously deposited stack of sediments in which the location of deposition — and therefore the nature of the sediments — can only change gradually.

One of the most important discoveries of sequence stratigraphy was the impact of sea-level changes on the morphology and stratigraphy of continental shelves. As sea level drops below the shelf edge, rivers erode into the shelf and incised valleys form. These valleys were thought to be erosional and accumulate little or no sediment during sea-level fall; they would only be filled during subsequent sea-level rise. If this was true, it would also mean that the erosional surface — or sequence boundary — at the base of the valley was a topographic surface at some point in the past, and all of the 'missing' sediment was transported beyond the shelf edge, into deeper water. But studies of incised valleys that formed during glacial periods of the Quaternary, and flume experiments in which both topography and stratigraphy can be carefully tracked have started to suggest a different, more complex story: overall fluvial incision can still be accompanied by deposition, the sequence boundary is a 'composite' surface that does not correspond to any topographic surface from the past, and significant parts of the valley fill have formed during incision. As the title of a paper in the *Journal of Sedimentary Research* suggested, an incised valley is a 'valley that never was.'

If this wasn't confusing — or, depending on your point of view, enlightening — enough, experimental stratigraphy has also cast doubts on one of the other tenets of sequence stratigraphy: the idea that all the deposits above the sequence boundary are younger than the deposits below. In a flume tank at Saint Anthony Falls Laboratory in Minneapolis, coastal deposits that were time-equivalent to fluvial sediments above the sequence boundary ended up lying below the erosional surface as the system slowly advanced into the miniature sea. This concept is not easy to visualize, but it turns out that one of the well-known small-scale sedimentary structures, which is mentioned in every sedimentology textbook, shows all the characteristics that seem so surprising in the case of large-scale erosional surfaces. Climbing ripples that have a relatively small angle of climb deposit cross-laminated sands with erosional surfaces that are clearly different from the topography of the ripples at any time; and it is tempting but foolish to think that all the sediment above one of these surfaces is younger than everything below.

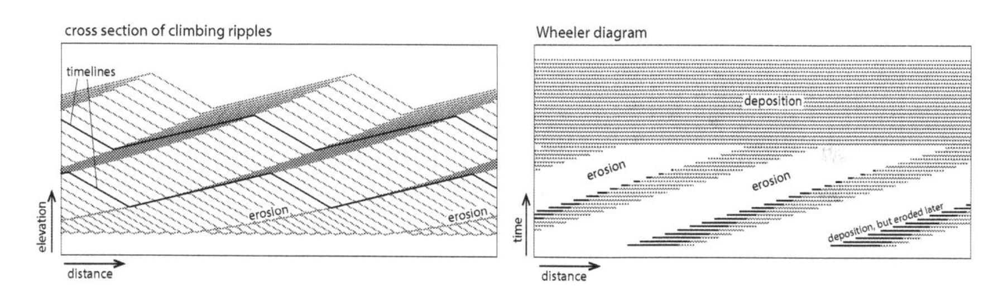

As geomorphologists and stratigraphers come up with new and more precise methods of age dating, as experimental sedimentologists and numerical modellers refine their techniques for modelling sedimentary systems in flume tanks and on powerful computers, and as field geologists and seismic interpreters adopt a more sophisticated view of stratigraphy, Harry Wheeler's ideas seem more relevant than ever. He initiated the revolution in our thinking about stratigraphic surfaces, but the revolution is not over yet.

References

Strong, N and C Paola (2008). Valleys That Never Were: Time Surfaces Versus Stratigraphic Surfaces. *Journal of Sedimentary Research* **78**, 579–593.

Wheeler, H (1964). Baselevel, lithosphere surface, and time-stratigraphy. *Geological Society of America Bulletin* **75**, 599–610.

Study modern analogs

Shahin Dashtgard

'I work on beaches,' is always a good conversation starter, and for many people it conjures up images of relaxing in the tropics. Although that's a pleasant thought, working in modern environments is rarely relaxing; rather it is hard work undertaken to bring clarity to sedimentary geology. Through modern research, geoscientists try to answer questions of what, where, why, and how and shed more light on the rock record. Modern research lets us access a wide range of depositional environments, and enables us to determine the conditions under which sediments are deposited and colonized by organisms. These data provide the fundamental basis for interpreting sedimentary strata, and hence form the basis for seeking out new resources (e.g. hydrocarbons and water) contained within sedimentary rocks.

Consider Waterside Beach in New Brunswick, Canada, pictured here in 2011 with a sediment grain size map made in 2004. It is a good example of a beach with a large tidal range, which enables close examination of sedimentary structures and burrows at low tide.

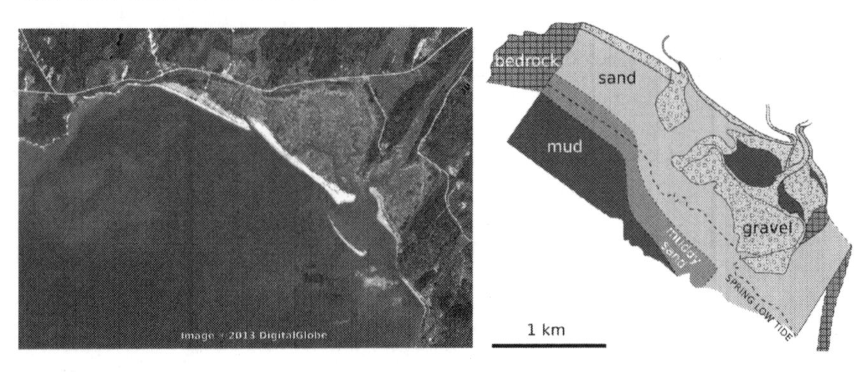

Waves breaking on the beach and shoreface produce a predictable set of sedimentary structures (sedimentology) that change as waves shoal onshore. As the sedimentary structures change, there is a corresponding change in the types of animals that colonize the sediment, leading to unique combinations of burrows (ichnology) constructed by those animals to cope with the environmental conditions. Depending upon the grain size of the sediments, the size of waves affecting the coastline, the sediment source, the geography of the basin with respect to its coastline, and the climate, astute geoscientists can determine the range of variability that can be found in the resulting depositional facies of beach–shoreface sediments. This integrated approach provides the basis for making effective rock-record interpretations, which enables us to better predict the architecture of sedimentary strata and determine the nature of the paleoenvironment. For petroleum geologists and hydrogeologists, such modern environment data constitute the analog models essential for exploring or exploiting resources contained in sedimentary rocks.

In addition to understanding the rock record, studying modern environments provides information on the plan view (lateral) distribution of sediments that cannot be accurately resolved from outcrop or subsurface geological studies. Walther's Law, indeed, is founded upon the realization that knowing the spatial distribution of environments and their deposits is critical to understanding the vertical succession of facies we see preserved in the rock record. This is of fundamental importance to sedimentary geology because most new resources contained in sedimentary rocks reside in the subsurface and can only be exploited through drilling. Whether developing an exploration strategy for finding new resources, building a development strategy for a newly discovered oil pool or aquifer, or trying to predict the lateral extent of permeability barriers (e.g. mudstone beds in sandstone), data derived from modern environments serve as the starting point for understanding the depositional system.

Modern environments have been studied for nearly as long as the science of geology has existed. As elsewhere, technological advances have enabled us to push the limits of what we can do. Nowadays, modern research takes place in environments from the top of the Himalayas to the bottom of the Marianas Trench, and is heavily reliant on advanced technology. This has led to a plethora of information on nearly all depositional environments on the planet. Yet with the increased use of computer modelling there is a need for more quantified data. Now instead of asking, 'Where is mud being deposited in estuaries?', the question has become, 'How thick and how laterally continuous are mud beds deposited at the bottom, in the middle, and at the top of an estuarine succession, and how does this change along its depositional profile?'.

Although new research questions are more tightly constrained, the fundamentals of modern environment research remain the same — to continue to build an understanding of the rock record. Modern analogs have been and remain a cornerstone of sedimentary geology.

Tectonic permeability

Fraser Keppie

Tectonic permeability is secondary permeability caused by plate tectonic processes. It is well-known that fractured rocks have an enhanced ability to transmit fluids and potentially to act as fluid reservoirs as well, especially if the fractures are exposed to critical stress conditions. Consequently, the density and orientation of fractures relative to the local stress field are important factors in any geological model. Usually, both the historical and present-day tectonic context of the basin control these properties, and this is why it can be so useful to evaluate the tectonic drivers of brittle deformation during hydrocarbon system analysis.

The first time I really appreciated the economic value of linking predictions of tectonic deformation to the hydrocarbon potential of a sedimentary basin was in a talk by Peter Hennings on the Suban Field of Sumatra, Indonesia (Hennings et al. 2012). He presented a multi-scale analysis of the tectonic processes governing fractures and stress conditions in the Suban Field. He presented how:

- At a regional scale, oblique relative plate motion vectors increase northwestwards across the Indo-Australia–Sundaland subduction zone causing components of strike-slip strain to increase in the upper Sundaland Plate.
- At a basin scale, partitioning of oblique strain across the Sumatran Fault, above the Indo-Australia–Sundaland subduction zone, produced dextral and compressive stresses in the Sumatran back-arc and thus in the Suban Field.
- At the local scale, how determining the orientation and relative magnitude of the maximum horizontal stress and its relation to local fractures from available well data allowed him and his team to evaluate the number of fractures at critical stress conditions.

As I remember the story, this analysis ultimately allowed his team to help plan the most economic well in the history of ConocoPhillips with a deviated path designed to intersect a maximum number of productive fractured zones.

More recently, I have had the opportunity to review the tectonic evolution of the Gulf of Mexico. In this setting, it is conspicuous how the inferred trace of the Tamaulipas–Golden Lane Fault in the western Gulf of Mexico basement (Pindell 1985) is coincident with the interpreted traces of the Faja de Oro fault

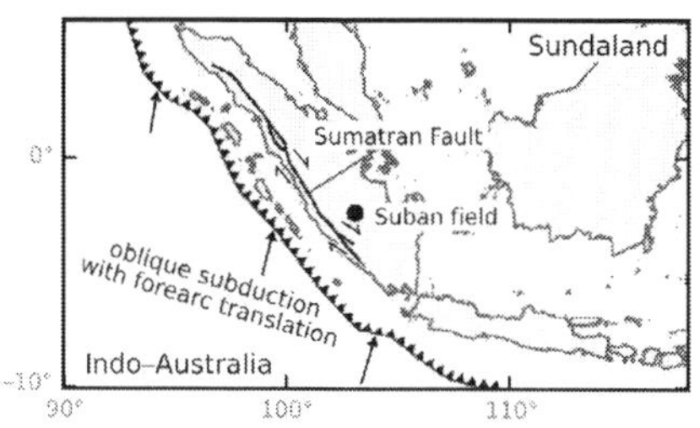

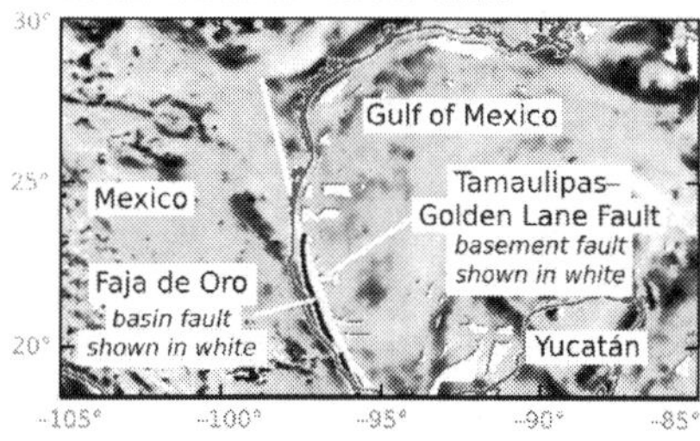

system in the overlying basin fill (le Roy et al. 2008), all of which underlie the region where some of the largest hydrocarbon resources have been recovered. It seems likely that re-activations of the basement fault may be directly related to faulting in the overlying basin and relative increases in both the tectonic permeability and hydrocarbon yield along this damaged zone.

These and other examples demonstrate the value of understanding the tectonic controls and deformation history of a prospective basin at multiple scales. The question for the hydrocarbon geologist is where else could a focus on tectonic permeability yield a competitive advantage? To assess this opportunity, a high-level workflow can be identified as follows (e.g. Hennings et al. 2012):

1. Identify a basin with a classic hydrocarbon system (i.e. where source rocks, reservoir rocks, traps, and thermal maturation are givens), but where a history of deformation is evident.
2. Reconstruct its tectonic evolution identifying, if possible, major faults and diffuse deformation zones.
3. Reconstruct its modern tectonic setting in which boundary conditions for its active state of stress can be hypothesized.
4. If possible, extend the structural and stress predictions to the third dimension using seismic and potential field data.
5. If possible, evaluate the local stresses on local fractures using down-hole well imagery and logs.

References

DeMets, C, R Gordon, and D Argus (2010). Geologically current plate motions. *Geophysical Journal International* **181**, 1–80.

Hennings, P, P Allwardt, P Paul, C Zahm, R Reid Jr, H Alley, R Kirschner, R Lee, and E Hough (2012). Relationship between fractures, fault zones, stress, and reservoir productivity in the Suban gas field, Sumatra, Indonesia. *AAPG Bulletin* **96**, 753–772.

Pindell, J (1985). Alleghenian reconstruction and subsequent evolution of the Gulf of Mexico, Bahamas, and Proto-Caribbean. *Tectonics* **4**, 1–39.

Le Roy, C, C Rangin, X Le Pichon, H Nguyen Thi Ngoc, L Andreani, and M Aranda-Garcia (2008). Neogene crustal shear zone along the western Gulf of Mexico margin and its implications for gravity sliding processes. Evidences from 2D and 3D multichannel seismic data. *Bulletin de la Societe Geologique de France* **179**, 175–193.

The Dark Art of regional geology

Tony Doré

A friend and mentor of mine delights in telling me, 'Tony, make no mistake: geology is a Dark Art!' He should know I guess, since this dubious practice has taken him to the upper echelons of the oil industry. Furthermore, many of you reading this will know immediately what he means. There's something a bit suspect, a bit smoke and mirrors about geology. It's far too imprecise to be a real science, right?

But what's so dark about it? Well, perhaps it's because you sometimes have to conjure something out of almost nothing. Perhaps it's because what you conjure up is kind of counterfeit, a bit suspicious, and very seldom the product you were expecting. Or perhaps it's because not everyone approves of the results of our art, which includes the black stuff that — for a lot of us — pays our wages. And let's not forget, we are the Dirt People, as theoretical physicist Sheldon Cooper haughtily declared in *The Big Bang Theory*.

Why an art? Well, geology wasn't originally a recognizable science at all, more the preserve of leisured aristocrats and clergymen who saw geology as an extension of theology. Even now, a disproportionate number of geologists seem to moonlight as painters, poets, or musicians. More than most sciences, geology is about imagination and creativity. It uses diverse media and initially unpromising materials to create a credible picture — pretty much the way an artist does. So yes, it's a bona fide Dark Art. And it's my contention that regional geology is the darkest art of all.

But wait, you say. Geology's not like that anymore. The solitary visionary trying to reconstruct ancient worlds with a hammer and a grubby map — that doesn't represent geology now. Our discipline has entered a new domain where it's far more precise, more analytical, more experimental. Or as I read somewhere a few years back, 'Geology has now moved out of the field and into the laboratory'.

Well, if that's true, I for one am going back to playing my guitar. But I suppose I know what they mean. Things have changed unrecognizably over the last few decades. Laboratory analyses, 3D seismic volumes, limitless processing power, and infinitely flexible workstation techniques have given us unprecedented scope for analysis and precision. That kind of precision is pretty important in

More than most sciences,

geology is about imagination and creativity.

the petroleum industry when it comes to pinpointing the remaining oil in a field, or guiding a horizontal well along a thin sandstone layer. A new breed of geologist has even arisen to ride this wave. I think of them as the Detail People.

I truly respect the Detail People. 'God is in the details,' we're often told. And how can you not admire someone who knows how to focus the technology on the tiniest forensic points, relentlessly interrogating them until they beg for mercy and finally yield their story? Not least, I love the Detail People because they can do something I can't. Maybe due to a chronic lack of patience, my own grasp of detail has never been wonderful.

However, there's one thing that many of the Detail People can't do, and that's envisage the big picture. In my current company role of roaming advisor and busybody I see this often. I might, for example, find myself sitting with an interpreter who is coaxing beautiful 3D renderings of depositional systems from a seismic volume, but who doesn't know — or particularly care — what happens just outside the block, let alone what the tectonic setting in the Late Cretaceous was.

This isn't a rare occurrence — it's almost the norm. Furthermore, it's not unreasonable. As you spread your net wider things get fuzzier, more hypothetical, the unique solutions disappear, and that makes people more reluctant to commit. It's time to summon a devotee of the Dark Art. This is their domain. It's a murky, uncertain place, but it's also where many big discoveries are made.

This essay continues in *The Dark Art's great payoff.*

Acknowledgments

Sheldon Cooper is quoted from *The Big Bang Theory* season 4, episode 15. "The Benefactor Factor" aired on 10 February 2011.

The Dark Art's great payoff

Tony Doré

The jigsaw puzzle is a very seductive proxy for any kind of mystery, mainly because you just have to determine which piece fits where and eventually the whole picture will gloriously emerge.

The trouble is, some vindictive person (probably that theoretical physicist I mention in *The Dark Art of regional geology*) has grabbed a handful of pieces from the middle and thrown them out of the window. They're going to take a long time to find! We still want to complete the picture — but how?

The highly focused scientists I call Detail People would say, 'Obviously, we need more data.' True, but right now we don't have that luxury. Fortunately, regional geologists are quite comfortable, because that hole is where they spend most of their time. To them, there's already an abundance of information. The hills and fields around the outside tell you it's a pastoral scene. It's reasonable to suggest more hills or a lake in the middle. A tractor or grazing cow is quite plausible. On the other hand, a battleship is vanishingly unlikely.

That's the regional geologist's job — to propose credible models using scattered and uneven data, and to extrapolate, often over long distances. You can confirm or modify this working model as you get more data. Of course, you might have been spectacularly wrong, too. In which case, open that window again and throw your hypothesis out. There's no room for pride in regional geology. Good regional geologists crave confirmation of their ideas, but they

The next big oil play
is already in our heads.

love arguing even more. The last thing they want is for their ideas to be slavishly applied without challenge.

Why is all this useful? The answer is that great new ideas often arise by combining the knowledge of the Detail People, who don't have the time or inclination to think about how their areas relate, into what regional geologists do. Good regional geologists beg, borrow, and steal. They promiscuously absorb other people's ideas and weld this second-hand information into a new whole. In the petroleum industry, that new whole can sometimes lead to big discoveries.

Major oil companies have vast databases, consistent methodologies for calculating risk and resources, and an ability to rank every basin on the globe. And yet they are not usually the first entrant into a new play. That's because, despite what pundits tell you, you can't systematize your way to frontier success. In fact, most of the major breakthroughs over the last decade, both conventional and unconventional, were made by independent oil companies without the benefit of big databases and systematics. Sure the big boys came in later, wallets bulging, to mop up the spoils, but for all their advantages they weren't the mould-breakers.

I have asked a number of experienced industry analysts why this is the case. Were the successful independents just the lottery winners? For every success, were there another 20 who went bust? Well, maybe. But it also turns out that a common factor between many of the successful independents was their hiring policy — the deliberate targeting of geologists with proven track records and big regional ideas. They couldn't afford to amass a global investment portfolio and play the odds, so instead they opted to dabble in the Dark Art. They brought in a few highly qualified obsessives, and gave them the resources to play their ideas out. Invariably, those ideas came down to making connections: extrapolating between basins, or boldly extending a geological idea into a new domain.

All of which might explain why I ended a recent talk to a group of regional geologists with the slogan 'the next big oil play is already in our heads.' I meant that the individuals in the room had bodies of knowledge which, when combined and nurtured, might result in a flash of insight and a new exploration direction. For me, seeing a new idea materialize in that way is one of our Dark Art's great payoffs.

The golden age of seismic interpretation

Paul de Groot

The European Association of Geoscientists and Engineers is one of the leading geoscientific organizations in the petroleum industry today. Each year thousands attend its annual conference and exhibition. At these events the majority of papers and exhibits deal with seismic acquisition, processing, and interpretation — indicating that seismic technology is experiencing a golden age.

Advances in seismic technology have indeed been phenomenal, especially in the domains of seismic acquisition and processing. Denser sampling in space and time, better algorithms to remove unwanted noise, and ever-improving imaging allow geoscientists today to construct increasingly accurate geological models on which important exploration and production decisions are based. Progress has also been made in the domain of seismic interpretation through the introduction of sophisticated attributes and innovative visualization techniques. However, it is my belief that in the field of seismic interpretation we are yet to enter the golden age. In fact the best is yet to come. Let's look at what has been achieved in seismic interpretation to date, and the range of future possibilities.

To unlock the hidden treasures in seismic data today we need to introduce the concept of geological age into the interpretation process. Time in geology is a poorly understood feature. Professor Salomon (Salle) Kroonenberg, an *éminence grise* of Dutch geology, wrote a very interesting book about the subject, *The Human Scale: The earth ten thousand years from now*. Salle says time in geology manifests itself in three ways: as a flow, as a pulse, and as a wave. For example radioactive decay, the expansion of the universe, and evolution are processes in which time manifests itself as a flow — what Stephen Jay Gould called time's arrow. Earthquakes, meteor showers, eruptions, floods, and extinctions are examples of time as a pulse. The third manifestation, time as a wave, is about cyclical processes. Examples are continental plates breaking apart, colliding, and breaking again; ice ages alternating with warmer periods; and sedimentation processes controlled by Milanković parameters.

In seismic interpretation we are primarily concerned with time as a wave. In the 1970s Pete Vail and his colleagues at Exxon started to map seismic reflection patterns. They observed cyclical patterns that could be explained in terms of sedimentation processes. Moreover, they discovered that seismic reflectors are the first order approximations of geological timelines. In other words, mapping horizons that follow seismic reflectors is essentially equivalent to mapping geological time. A new interpretation technique called seismic sequence stratigraphy was thus born. The technique helped improve our understanding of depositional systems and has been used ever since to find stratigraphic traps. Due to the lack of supporting software algorithms, seismic sequence stratigraphy, however, never reached its full potential with interpretations carried out manually in what was a cumbersome and time-consuming process.

Recently, however, a new group of semi-automated seismic interpretation techniques have emerged that aim to generate fully interpreted seismic volumes. The algorithms behind these techniques all share in common the fact that they try to correlate seismic positions along geological timelines.

- Tracy Stark's 'age' volume assigns a value representing relative geological time to each seismic sample position. The age assignment is based on correlating instantaneous phase signals from trace to trace.
- The PaleoScan software from French startup company Eliis builds a geological model roughly on the scale of the seismic sampling by connecting each seismic event to the most probable neighbouring event.
- Volumetric flattening, Chevron's proprietary technology by Jesse Lomask, uses similarity-correlated surfaces to flatten the seismic volume or attributes of it. The flattened volumes are known as Wheeler cubes.
- dGB's HorizonCube algorithm correlates timelines in the pre-calculated seismic dip field. The tracked surfaces are stored as a dense set of mapped horizons called HorizonCube.

Fully interpreted volumes are used, among other things, to assist in well correlations, in unravelling depositional histories, and finding stratigraphic traps using sequence stratigraphic interpretation principles. Other applications include detailed geological model building and improved seismic inversion and reservoir property prediction schemes that start from more accurate low frequency models. In addition, geohazard interpretation, geo-steering, and the finding of sweet spots in unconventional plays have all benefitted significantly from fully interpreted volumes that can add value to the seismic data.

While enormous progress has been made, it's clear that the current group of global interpretation techniques is still evolving. Interpretations covering different geological settings such as passive margins and carbonate platforms have already been published. But as the technology matures, additional applications and more complex settings will be put to the test. What is clear is that we are on the verge of entering a golden age of seismic interpretation. Watch this space!

The topography of tectonics

Tim Redfield

Geologists make erratic drivers. There are just too many interesting things to watch out for along the side of the road: faults, folds, and the details of sedimentary layers, to name just a few. But most geologists are also in the habit of stopping their car from time to time at a roadside scenic view in order to take in the sweep of the landscape, particularly its topography. A lot of geological information is embedded in topography, and this is especially true of tectonics.

Tectonics can be loosely defined as the set of processes that deform the rocks of the earth's crust. For example, these can include the motion of the large lithospheric plates to which our continents belong. It is an easy thing to imagine how a slow collision between India and Asia caused enormous sheets of rocks to be thrust over or under one another. The Himalaya mountain chain is our textbook example of tectonic topography.

That's not all to the story, of course. There is a tremendous, ongoing argument over the relative importance of tectonics versus climate and erosion in the Himalaya. One camp is adamant that once the mountains became sufficiently high they forced prevailing winds to flow over them. By cooling the air the mountains caused the annual monsoon rains, which in turn created increased erosion and more isostatic uplift to compensate for all the material washed out to sea. This is certainly a good point, but the opposing camp has an instant rejoinder. 'Yes,' says the tectonicist, 'but without tectonic forces to have raised the earth's surface, no mountains would have existed to make the monsoon in the first place!'

Mountains such as the Himalaya, the Alps, and the Andes are spectacular. But the earth's topographic envelope also records the effects of less grand but equally important tectonic processes. A photograph that appears in many geology textbooks shows a concrete sidewalk curb in Hollister, California which has been broken and offset by the slow, steady, sideways creep of the Calaveras fault. It is easy to take a step back and imagine how active faulting can rearrange an entire landscape. Just like the sidewalk, a creek or a river that crosses an active fault will have to adjust itself after a large earthquake. Surface rupture can cause a displacement of many metres, which means its effect on a landscape can be rather substantial. One great earthquake that occurred in Owens

Geologists make erratic drivers. There are just too many interesting things to watch out for along the side of the road.

Valley, California in 1872 created a fault scarp almost 10 metres high. California's Sierra Nevada is on the side that went up. Repeated over and over for millions of years, this kind of faulting will make mountains.

No fault can run forever: all faults must have beginning and end points, which we call the fault tips. Offset will be greatest near the middle of the fault, and there the fault scarp will be at its highest. The scarp height will decay parallel to the fault plane in both directions, becoming zero at each fault tip. The scarp face will be steep and slope towards the fault, while the backside will slope gently away from the fault. This is called asymmetric escarpment topography and is quite common in regions that are undergoing active extension. It is also common on the continental scale at what are (probably incorrectly) called passive margins.

Folds in the earth's crust can also influence topography. Imagine a set of sedimentary layers, one on top of the other and some better compacted and more able to resist erosion. Next, imagine the earth squeezing them, not enough to make thrust or reverse faults but sufficient to warp them upward into an elongated dome. Finally, imagine this structure — called an antiform, or sometimes an anticline — is still being squeezed, and is thus actively growing. In becoming higher its footprint gets larger: as the nose of the antiform migrates horizontally it can push rivers out of the way like a bulldozer making a dam. But should the softer core of our antiform become breached by erosion, the inner layers may be washed away very rapidly. The resulting landscape may end up looking like an amphitheater with some very diagnostic river patterns around and inside it.

With the proper geological toolbox a lot of tectonic history can be recovered from topography. These few patterns we've discussed constitute only a partial list of what causes the tectonic topographer to behave unpredictably behind the wheel. More formally, we should sub-categorize our traffic hazard as a tectonic geomorphologist. Tectonic geomorphology infuses the study of landforms with knowledge from sedimentology, geochronology, seismology, structural geology, and many other types of geology. Specialists in these and other fields commonly work together, one waxing poetic and waving his arms in the air whilst the other ignores him and tries to safely navigate the road ahead.

The trouble with seeing

Evan Bianco

You've probably heard the adage, 'the best geologist is the one who has seen the most rocks.' I have grown to dislike this statement. It's not that there is something false or incorrect about it per se, but it's often served as a trump card severing any deeper conversation about skills and performance. It's too trivial, and it's doing us all a disservice.

What's left unsaid, and therefore goes unappreciated, is that *seeing* is not just another measurement or experiment we do in earth science. Rather, it is a learned process of perception that draws from two distinct skills: recognition and cognition (Dell'Aversana 2013). Recognition is the low level sensory activity of detecting patterns and signals. Cognition, on the other hand, is the higher order faculty of processing and negotiating with information, guided and biased by experience, theoretical knowledge, and heuristics. If you aren't practised at both, you won't *see* any rocks at all.

Furthermore, in this light, the concept of a geological dataset becomes puzzling. There is no data as such. Within a piece of core, or a seismic line, are objects that are perceived in different ways depending on the observer's background. Our perception depends on the theoretical knowledge that we have about that object. A person without knowledge does not perceive an anomalous signal in the same way as an expert. Two experts will see different signals according to their individual recognition and cognitive abilities. The implication is that observations, which often pass for data in geology, are inherently tied to the creator, subjective and tacit. It means we can't perform any sharp distinction between data and models, or observation and interpretation. It's no wonder two geologists may arrive at divergent interpretations, they don't even start with the same inputs.

This is troubling because geologists work in teams and solve problems of economic, social, and scientific importance. The problem of perception has a detrimental effect on our experiments in two key respects: we try to measure what we are looking for, and we see what we want to see or have been trained to see (Bond 2012). The reality of working with others means reconciling differences in perception. Here are a few suggestions to cope:

Observations, which often pass for data in geology, are inherently tied to the creator, subjective and tacit.

Show people how you perceive. Show them both pieces, the cognitive and the sensory components. Draw attention to your own ingrained practice so you can modulate it from within. In a way, you are disclosing your scientific method, allowing others to make their own connection.

Documents hold up better than hearsay or anecdotes. Think of all the stuff you produce at intermediate steps of geological work; drawings, sections, geological maps, descriptions, and so on. They are embodiments of knowledge, and they often form a readable arc of your mental path within a geological investigation.

You aren't a politician. You are allowed to change your mind. Going over your process again and again enables self-criticism and exposes holes in your data.

Search for coherency and consistency between different clusters of information. Instead of thinking in a linear fashion from data to model, or observation to interpretation, consider each as a lens pointing to a complex earth system expressed at varying levels of abstraction.

Beware of vocabulary and jargon. With the exception of digits that come from a logging tool, our vocabulary for describing geological objects may be inadequate for your needs. Is that contour map a model or is it data? In the end, it may not matter what you call it, but it is at the risk of misuse if it is taken out of context.

A geologist that is strong on theory but poor on detection won't be able to solve practical problems. One that is weak on theory but great at detection is behaving more like a camera. The weaving of both is where you add value, playing both the detective and the detector.

References

Bond, C (2012). Recognize conceptual uncertainty and bias. In: *52 Things You Should Know About Geophysics.* Agile Libre. 132 p.

Dell'Aversana, P (2013). *Cognition in Geosciences: The feeding loop between geo-disciplines, cognitive sciences and epistemology.* EAGE Publications.

The world is your laboratory

John Harper

I am a geoscientist. My career is my passion; it is not just a job. I have seen geology around the world, from the oldest rocks on earth to the youngest sediments alive with life. I have learned many useful lessons about my career, my relationships, and my life. It has been an incredible journey, one I could not have anticipated in the beginning.

As a young lad the outdoors, the forests, and the animals were a great wonder to me. I would spend whole days alone wandering, soaking in all I could learn. I wanted to be a wildlife officer. In preparation I spent summers as a junior ranger with the Ontario Department of Lands and Forests. Though I love the outdoors, I decided I didn't want to live — or raise a family — in small, isolated towns. My whole perspective changed one career day in high school. An oil geologist from Calgary gave a presentation on exploration. He spoke of summers in the field and horseback riding through Jasper National Park. That did it! Geology was for me.

I was an average student at university. My summers were spent in the field as a hydro lineman, a miner in Elliott Lake, and most significantly as a field geologist for the Geological Survey of Canada's Roads to Resources program. For several months at a time my partner and I moved north by canoe, portaging to new campsites every four days. We undertook field traverses, two downstream and two upstream of each of our camps, received food by air once a week if the plane didn't lose track of us, and met aboriginal peoples who had never been out of the bush. I came out of this experience with new certainty that I did not want to live in the bush all my life. So I changed direction.

I took a summer teaching certificate and taught high school for a year. I enjoyed the children but I found that I became too involved in their problems, which were many and in some instances horrendous. I realized that if I was going to teach I would have to teach the subject I loved. That required that I go to graduate school.

At this point I met two key individuals in my life: my girlfriend at the time — now my wife — and Professor Frank Beales of the University of Toronto. The faith and support of both these individuals allowed me to prove I had what

Be aware that it is you who has to do the network maintenance: don't wait for others to contact you.

it would take. I accepted the challenge of the master's degree and immersed myself. I made it my goal to understand my subject from the very basics to the advanced theories. I intended to be the highest calibre geologist I could be.

After then obtaining my PhD, I joined Shell Development Research which gave me tremendous opportunities to grow scientifically and intellectually. I worked on geological problems throughout North America, from the depths of the Michigan Basin reef play to the regional development of the Appalachians, the Arctic, and Canada's East Coast. Since then I have consulted in the Philippines, Fiji, Libya, and the US, taught at the Centre for Earth Resources Research at Memorial University of Newfoundland, and worked at ConocoPhillips and the Geological Survey of Canada — what a thrill to work with such esteemed scientists.

What have I learned on my fortunate journey?

- I have learned to not leave my career in the hands of others. No one knows better than yourself the scope of your capabilities and for sure no one can advocate for you better than yourself.
- Technical quality and capability establish your reputation.
- Seek broad integrated skill sets such as seismic, structural geology, paleontology, well-log analysis, rock properties, and so on.
- Always have at least two career directions available to you at the same time.
- Develop a network of like-minded colleagues and maintain the relationships. Be aware that it is you who has to do the network maintenance: don't wait for others to contact you.
- Build your confidence by accepting challenges and broadening your global experience. To be a fine geologist you need to develop as broad an appreciation of how earth processes work as you can.
- Be prepared to teach them and be willing to mentor younger geoscientists.

I could not have asked for a better career and hobby. I wish you luck and fulfillment in yours.

Three kinds of uncertainty

Weishan Ren

A good quantification of geological uncertainty is demanded by almost every hydrocarbon development project. Uncertainty is normally presented as part of key project parameters, such as original hydrocarbons in place, recovery factor, and net present value. Petrophysical properties such as porosity, fluid saturation, and permeability are directly used in calculations of these economic indicators. Geology — in terms of facies proportions, geobody size, and geological trends — may not be directly involved in the calculations, but is implied in the distribution of petrophysical properties. Therefore, geological uncertainty, which is the uncertainty in facies proportions and so on, should align closely with the uncertainty calculated from petrophysical property models. And conversely, our geological model — lithofacies distribution in particular — should not only be geologically sound, but also be adequately described by its petrophysical properties.

Quantifying geological uncertainty requires both a good geological understanding and advanced geostatistical methods. Our geological knowledge and experiences help us considerably in data interpretation and constructing conceptual depositional models. With sufficient wells and seismic information, we normally have good confidence in determining sedimentary environment, facies distribution, and sequence stratigraphic evolution. However, when we transfer our geological understanding from vertical wells to three dimensions, a lot of variability enters our interpretation. Between wells we have to deal with great uncertainty. Facies proportions, size and shape of geobodies, and geological trends are significantly influenced by different geological scenarios. How can we deal with all these sources and types of uncertainty? This is where advanced geostatistical methods come in.

Deterministic models — whose results are uniquely defined by their inputs — cannot reflect our knowledge about where and how much we are uncertain, so we must use probabilistic models. With these practical tools we can construct reservoir models by statistical inference from available data. Multiple sources of data, including core and log data, seismic attributes, regional geological trends, and production data, can be integrated by using multivariate geostatistical methods. The resulting heterogeneous reservoir model will honour input well

Quantifying geological uncertainty requires both a good geological understanding and advanced geostatistical methods.

data and their statistics, especially their spatial correlations. The uncertainty is presented in multiple realizations, or equally likely versions of the output.

To obtain a full characterization of geological uncertainty, there are three key sources of uncertainty you should know about:

1. Commercial geomodelling software packages such as RMS, Petrel, and GOCAD take care of well-data distributions, spatial correlations, trends, and random simulation paths, and provide geological uncertainty from the models. Most people use this uncertainty range to represent the full space of geological uncertainty; however, there are two big pieces missing.
2. The second key source is the uncertainty in our geological scenarios. Different scenarios significantly affect our gridding and geomodelling methodology and therefore the final uncertainty range.
3. The third is the uncertainty in input statistics such as the global mean values or percentages of different facies over our area of interest. These are generally treated as constant values in geomodelling, but observations show they change when adding new wells. Varying the mean values can greatly extend the uncertainty range.

Only a hierarchical modelling approach that accounts for all of these sources of uncertainty should be used to obtain a reasonable geological uncertainty range. Otherwise you're much less certain than you think you are.

Time and motion

Stuart Burley

The one topic that distinguishes geology from all other sciences from which we regularly steal and borrow, is time. It's the metric that enables much of what we see and postulate as geologists to be reality, but which we tend to take for granted. Indeed, it is commonly hidden from the casual observer. There are of course catastrophic events that take place geologically instantaneously and leave their mark in the geological record, but time is the differentiator that enables most geology to happen.

Why is time so important?

We know the age of the earth, with reasonable certainty and accuracy. It's around 4.7 billion years old, as determined from the decay of radioactive elements. We've come a long way since Bishop Ussher, in 1654, declared with naive precision that the earth was created on 22 October 4004 BCE, according to his analysis of the scriptures back through the generations of prophets. But it was Arthur Holmes, measuring the half-life of radioactive elements, who turned dating into a precise science and gave us quantifiable deep time. In 1913 he published *The Age of the Earth* and calculated its age to be remarkably close to the value accepted today. It is this time that allows geologically slow cumulative processes to have enormous effects and cover global distances. Without deep time, evolution, plate tectonics, continents morphed beyond recognition, oceans vanished, species lost in mass extinctions, hot house, ice house, and sea-level change would not be possible.

Include time in your thinking

Since the dawn of our understanding of plate tectonics in the 1970s, we have become accepting of large-scale lateral tectonic movements. Recognition of vertical movements of the crust have taken something of a back seat, despite well-accepted notions of plate collision, orogeny, and thermal relaxation subsidence. When we work in sedimentary basins, well data and seismic usually provide unequivocal evidence of both geological time and vertical movements. Take a look at the geoseismic section opposite. It doesn't matter where this section originates, suffice to say it's a tilted and partially inverted rift basin in which sequences and horizons are exceptionally well dated. The fault blocks at the northern end of the section are also present in the deeper southern part of the basin, but are here buried beneath a series of sequences that unconformably overlies the fault blocks and thickens to the north, the direction from which they were derived. The large inversion structure to the south is still active today, and sediments within hanging walls of the main fault blocks are thicker than in corresponding foot walls. Dating of the sequences indicates that significant amounts of time are represented by the interpreted bounding surfaces.

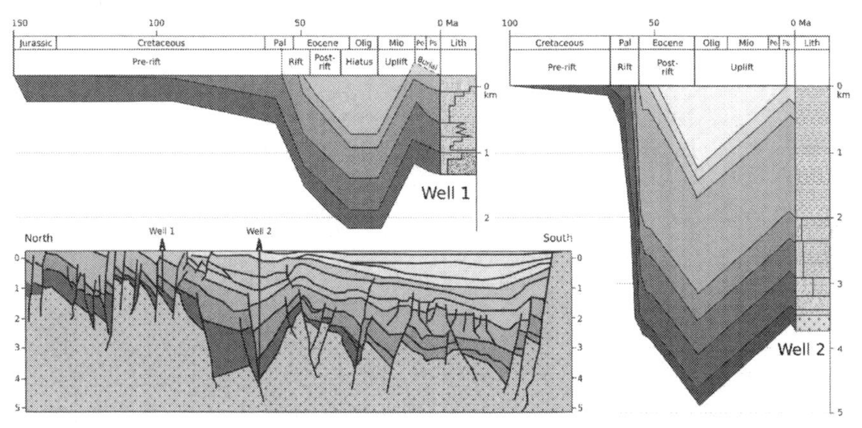

Visualize time and motion

A powerful visual way of highlighting changes through time in sedimentary basins is through one-dimensional burial history plots. These can be hand drawn, although there are many commercial packages that use sediment thickness data, sediment and surface ages, and knowledge (or guestimates) of uplift amounts. The resulting models trace the movement of sedimentary units or events in basins through geological time. Two models are located on the line.

Such burial models illustrate the remarkable variation in sediment thickness, erosion, and uplift across the section — in a quantitative time framework. Note the varying amount of uplift, and consequential time gaps in the section as illustrated by the two models. Next time you present a geological interpretation, supported by maps and sections in space, ask yourself how you can also traverse geological time and what vertical movements have taken place. Don't take time or motion for granted.

References

Aubry, M P (2009). Thinking of deep time. *Stratigraphy* **6**, 93–99.

Holmes, A (1913). *The Age of the Earth*. Harper, London. 228 p.

Lewis, C (2000). *The Dating Game: One Man's Search for the Age of the Earth*, Cambridge University Press. 216 p.

Trust your observation skills

Elisabeth Kosters

Becoming a scientist means that you must develop observation skills. With rare exceptions, observation skills are not taught explicitly either in secondary school or university. Most teachers spend little or no time on the importance of careful and methodical observation, on what that actually entails, on how existing knowledge and preconceived notions influence our observations, or on the trapdoor that lies between observation and interpretation. Students are generally only given instruction in how to record observations, usually in the context of a lab in which those recording methods are practised.

For example, in my first year, I had to draw the exquisite fossil specimens of our university's historical geology collection in a sketchbook, to be handed in for a mark. Instruction: draw the specimens in such a way that all the different body parts are clearly visible and label them. I did realize that this lab helped me memorize the fossils for the inevitable lab test, but that drawing taught me observation was lost on me.

I wasn't introduced to the idea that 'what you don't see, doesn't exist' until I was a graduate student. A charismatic professor casually introduced the notion as I was trying to record and understand bewildering sedimentary structures on a tidal flat. It was a revelation and I have been very conscious of it ever since. I think this awareness helped me improve my observation and recording skills in subsequent years.

At some point I had learned to observe the essential features and to record these using accepted codes. That meant that my observations could be checked and replicated, a crucial ingredient of scientific practice. I loved spending long hours in splendid isolation in a core lab, quietly working my way through metres and metres of core, slowly and carefully logging my way up through time. I learned to distance myself from interpreting, instead focusing on observing all features and letting the story tell itself. I loved it.

In the late 1980s I taught for a few years at the Institute of Earth Sciences of Utrecht University. Several faculty members had just participated in a scientific expedition in the seas around Indonesia. One of their sampling areas was Kau Bay, a dysoxic silled basin within the island of Halmahera, from where they had

I learned to distance myself from interpreting, instead focusing on observing all features and letting the story tell itself.

retrieved a number of cores. The sediments of Kau Bay are mostly extremely fine grained. All cores had been logged lithologically, but the paleoceanographers and geochemists were in need of more detailed insight of the sedimentological processes that filled the basin throughout the Holocene.

We picked a crucial 10 metre core, stored in 1 metre sections. We subjected the entire core to X-ray radiography, analyzed grain sizes, and painstakingly compiled everything in a log, one for each metre section. Anticipation was high when we put the 10 individual sheets together.

And then… no story. The log didn't tell us anything, there was no pattern, no trends. I refused to believe that this was the reality and asked about how the core sections were numbered. Sure enough, what I thought was the top metre was really the bottom one, the second from the top was the second from the bottom, etc. A misunderstanding — because I had made sure to ask beforehand what the labelling convention was.

We took the individual logs apart and taped them together in the proper order. The story instantly presented itself to us: three distinct facies were distinguished and they reflected clearly the different processes that were responsible for filling Kau Bay under changing climatic conditions. It was not only a huge relief, but also a terrific life lesson: I could trust my own observations.

References

Barmawidjaja, D, A de Jong, K van der Borg, W van der Kaars, W van der Linden, and W Zachariasse (1989). The timing of the postglacial marine invasion of Kau Bay, Halmahera, Indonesia. *Radiocarbon* **31**, 948–956.

Kosters, E, W Zachariasse, and D Barmawidjaja (1989). Sea-level controlled (?) facies variability in deep-marine Holocene sediments of a dysoxic sea, Kau Bay, Indonesia. *Annual Meeting British Sedimentological Research Group.*

Watch out for Paul Bunyan

Patrick Walsh

'Take a look at this prospect.' These few words carry great weight, initiating a process that can eventually cost a company millions.

As geologists, we are required to be optimistic or we would never develop anything. On the other hand, we need to be critical, hunting down and interpreting every last piece of data to construct a realistic geological model. I don't think I used to be conscious of this, but I've realized that I begin analyzing prospects by assuming there are major flaws in any previous assessments. Then whether or not I come to the original author's conclusion, I try to let my assessment sit and stew for a day or two before sending it up the chain.

As a manager, I would ask that you do the same. I have seen geologists, new and experienced, fall into the trap of believing other people's BS. This habit carries the greatest danger in marginal prospects because excellent and terrible prospects should be more obvious. We will use our geological model and a quantification of its uncertainties to rank the new prospect relative to ones we already have or others that are available. Our recommendations will be used to decide which leases to acquire and where to spend our exploration and drilling budgets. Team members must ask challenging questions and think of alternative models that also fit the data. We should also try to tease out the technical flaws, in a friendly way, whether they under- or over-estimate. Maybe there are zones behind casing that could be perforated; maybe there has been enough exploration to justify walking away; maybe new technology could improve performance.

Another dangerous temptation lurks in old prospects, places various companies have explored over time, with or without drilling. The data are sparse or scattered, the geophysics might be old, but there are rumours, perhaps even a mythology, of missed greatness. This is the one that got away, and we have an opportunity to reel it in. The reports are positive and with limited available data we may be swayed to give it a higher ranking or to believe the rumours. I have started referring to these as Paul Bunyan stories. Paul Bunyan is a folklore character, a giant lumberjack, whose exploits were exaggerated from stories of real pioneers. In exploration prospect terms, a Paul Bunyan is a place someone did some initial work and maybe drilled a well. The stories got distorted over time,

Data hunting is not glamorous, but finding missed information is crucial and is a great feeling.

via some real-world telephone game, and inflated hints of resource to gushers that collapsed. When you look hard at these tall tales, there are indications of resource, but the data are incomplete, maybe because exploration happened a long time ago or for some other reason. An optimistic geologist working for the selling company has written a glowing description. They do not appear to have any more data.

There is always more data out there. Find it, talk to the original geologists if possible. Find out what they were thinking. Data hunting is not glamorous, but finding missed information is crucial and is a great feeling. But most importantly, it can save thousands on new data and possibly millions on wells. Creating a critical interpretation of the available data, and ignoring the Paul Bunyan stories, is invaluable — and it is what we are paid to do.

What are turbidity currents?

Zane Jobe

Turbidity currents sculpt the seafloor into spectacular submarine channels, and are important conveyors of sediment from the continents into the deep ocean. Turbidity currents are so named because the current is a turbulent cloud of sediment flowing downslope under the influence of gravity (envisage a muddy avalanche). These flows are unpredictable and destructive, attaining speeds of up to 20 m/s (72 km/h) and flowing for hundreds of kilometres. Geologists study these events and their deposits in order to understand the processes that are responsible for the distribution of terrigenous sediment in the deep ocean as well as for natural hazard and resource prediction. A few spectacular examples of seafloor morphologies and historically observed turbidity currents are summarized below.

On 18 November 1929 an earthquake-generated turbidity current broke multiple submarine telecommunication cables south of Newfoundland, Canada (Heezen and Ewing 1952). Using the cable break times and seafloor mapping, the current thickness was estimated at 150 to 300 m thick and its speed was calculated to be 18 m/s (64 km/h). The turbidity current formed 2–5 m high gravel dunes and carried 100–200 km^3 of sediment onto the abyssal plain (Piper et al. 1988).

Another turbidity current, this one human-induced, occurred on 16 October 1979 off the coast of Nice, France. Landfill operations for the Nice airport runway extension oversteepened the slope in the Var submarine canyon and caused a submarine landslide, which evolved into a turbidity current that travelled for more than 120 km, breaking cables on the way (Mulder et al. 1997). The current swept a bulldozer and pieces of the airport embankment more than 15 km from the airport, and unfortunately resulted in the deaths of numerous airport and construction workers.

Turbidity currents like these erode and deposit sediment on the seafloor, forming geomorphic features similar in geometry and scale to those found on earth's subaerial surface. A beautiful example of this is the Bengal submarine channel–levee system, offshore east India (Kolla et al. 2012). This channel, fed by the Ganges–Brahmaputra river system, is more than 1000 km long and is highly sinuous. In cross section, the Bengal channel is more than 1 km wide and highly

Turbidity currents erode and deposit sediment on the seafloor, forming geomorphic features similar in geometry and scale to those found on earth's subaerial surface.

aggradational. Using seismic reflection data to cut a horizontal plane through the channel system, lateral migration scroll bars and oxbow cut-offs form the majority of the stratigraphy. These features are very similar in geometry and scale to large, continental-scale river systems — just look around in Google Earth.

With ever-increasing quality and resolution of bathymetric, seismic, and flow-monitoring data, we will continue to advance the knowledge of submarine channel and fan systems and the turbidity currents that mould them.

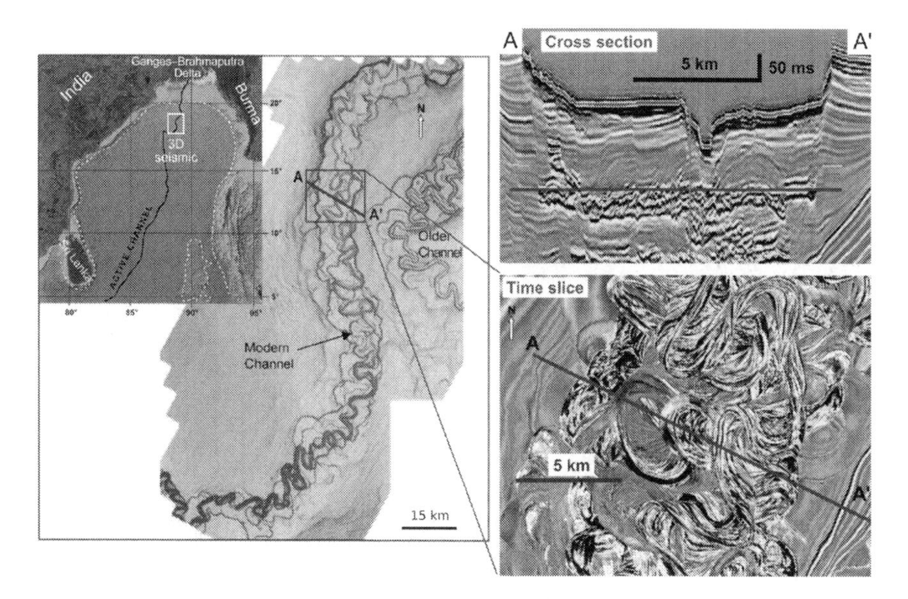

References

Kolla, V, A Bandyopadhyay, P Gupta, B Mukherjee, and D Ramana (2012). Morphology and internal structure of a recent Upper Bengal Fan-Valley Complex, SEPM *Special Publication* **99**, 347–369. The excerpted figure is copyright of SEPM.

Mulder, T, B Savoye, and J Syvitski (1997). Numerical modeling of a mid-sized gravity flow: the 1979 Nice turbidity current (dynamics, processes, sediment budget, and seafloor impact). *Sedimentology* **44**, 305–326.

Piper, D, A Shor, and J Hughes Clarke (1988). The 1929 Grand Banks earthquake, slump, and turbidity current. *GSA Special Paper* **229**, 77–92.

Piper, D and B Savoye (1993). Processes of late Quaternary turbidity current flow and deposition on the Var deep-sea fan, north-west Mediterranean Sea. *Sedimentology* **40**, 557–583.

What is a geologist?

Jim Barclay

Well, what is a geologist? I recall — and like — a definition I read once, something like: a geologist is 'one who studies the earth.' It's an all-encompassing definition, but it brings to mind the key aspect of what we do as geologists or geoscientists: we observe, we study the earth.

I like that this definition does not constrain who a geologist must be — there is no mention of university degrees, accreditations, experience, and so on. It does not specify the exact discipline within the field of geology — one could be a geophysicist, an engineer, a hydrologist, a geotechnical practitioner, and so on. One could be working in any field such as oil and gas, mining, water, aggregates, construction, archeology, or a world of others. The definition does not dictate very much in fact, other than the necessity of studying the earth.

The early geologists

The definition reminds me of the early naturalists of the 17th, 18th, and 19th centuries. I think of people like William Smith (1769–1839), the canal builder, who developed the concept of stratigraphy from looking at rocks all over Britain. The results of Smith's work — his 1815 Strata map — is shown on the facing page. And Nicolas Steno (1638–1686), a geologist turned Catholic bishop, who compared modern shark teeth to objects found in rocks and considered the objects to be ancient shark teeth. What unified such early geologists was their abundant curiosity, their use of keen observation, and their development of insightful interpretations. And they did so, remarkably, without the rich heritage of scientific context that we enjoy today. Indeed many did so against formidable cultural obstacles.

My theme here is to point out that geology is an unusual science, a visually driven observational science. For the most part, it is not dominated by mathematics, calculations, experiments, and such. Geologists do collaborate with many disciplines, however, and some geo-work falls fully within other areas. The geosciences are distinguished from other sciences by the level of interpretation required to deal with limited data and the fourth dimension of time.

What unified [the] early geologists was their abundant curiosity, their use of keen observations, and their development of insightful interpretations.

Art or science?

Does this level of interpretation make geology an art? Not really, but it perhaps inhabits a space somewhere between art and rigorous, hard science. It is the combination of observed data and an artful interpretation of typically incomplete data and an inability to be sure of what exactly happened in geological history.

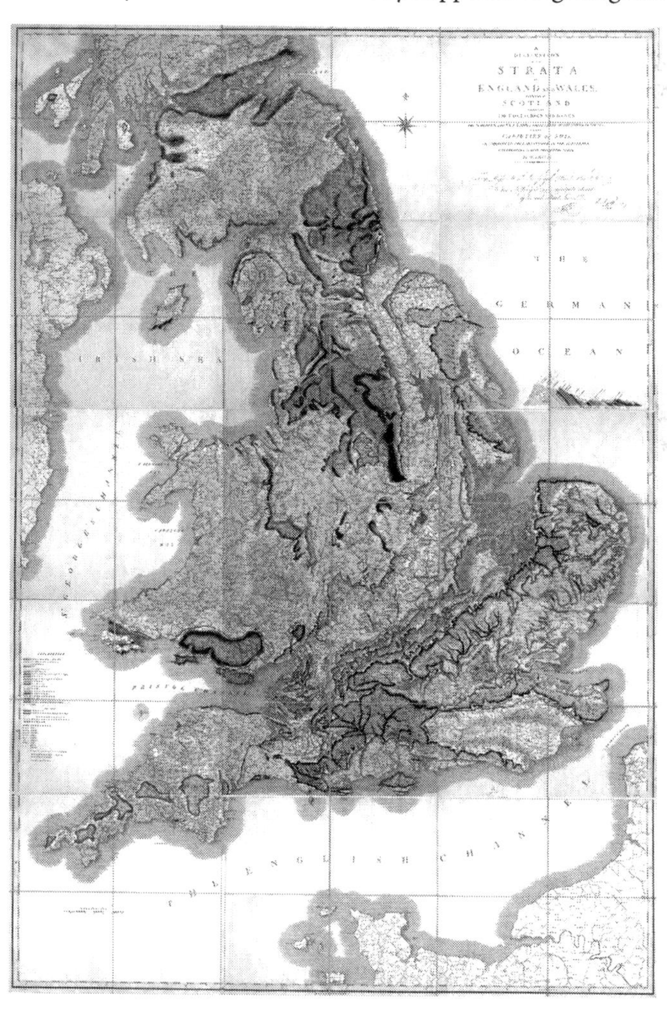

What is a great geologist?

Jim Barclay

Geology is an observational discipline (see *What is a geologist?*). But closely tied to the observational demands of the science, are the demands of curiosity. Curiosity is what drives the interest in observing. One of the greatest mistakes a geologist can make is the subjective viewing of the earth from a model-driven viewpoint. A model-driven approach is a kind of circular reasoning and ignores the value of evaluating multiple hypotheses. Models are useful. They give us a framework to gather observations (data) and build evidence to develop an interpretation(s). But models can blind us to important observations that maybe aren't consistent with the model's assumptions or somehow don't fit the model. And since we have trouble seeing what we are not looking for, we must depend on curiosity to uncover the truth. And we hope that curiosity can help us to see something we don't expect to see.

Geologists do work with conceptual models a lot — we must, given the infinity of unknowns about what happened in geological time. The trick is not to be fooled by them. I think of the quip that an engineer is one who thinks numbers are facts, whereas we geologists think that concepts are facts. Certainly, when working well, we show an ability to relate seemingly unrelated concepts (or facts!) and use them to create and bolster a story about the past — an interpretation. An ingenious example is the application of the observation of Saharan winds blowing dust westwards, far out into the Atlantic Ocean — then applying that model to western Canada Triassic rocks (Davies 1997).

According to Davies, the distribution of Quaternary dune-sand-derived turbidites off northwestern Africa, west-central Africa, eastern Brazil, and northwestern Australia has a relationship to offshore-directed winds transporting sand from coastal desert dune fields. The position of the present-day Saharan jet stream and related aeolian sediment transport is also shown in the figure opposite. Davies argued that similar processes, but in much shallower, lower-relief basin settings, may have occurred in the Lower Triassic Montney Formation of western Canada and in other Triassic units in that area.

While we use all the above talents in search of petroleum, how does a subsurface geologist work? I think dogged persistence and getting all the data possible go a long way towards successful prospecting — or 'sleuthing' as my friend and

Since we have trouble seeing what we are not looking for,
we must depend on curiosity to uncover the truth.

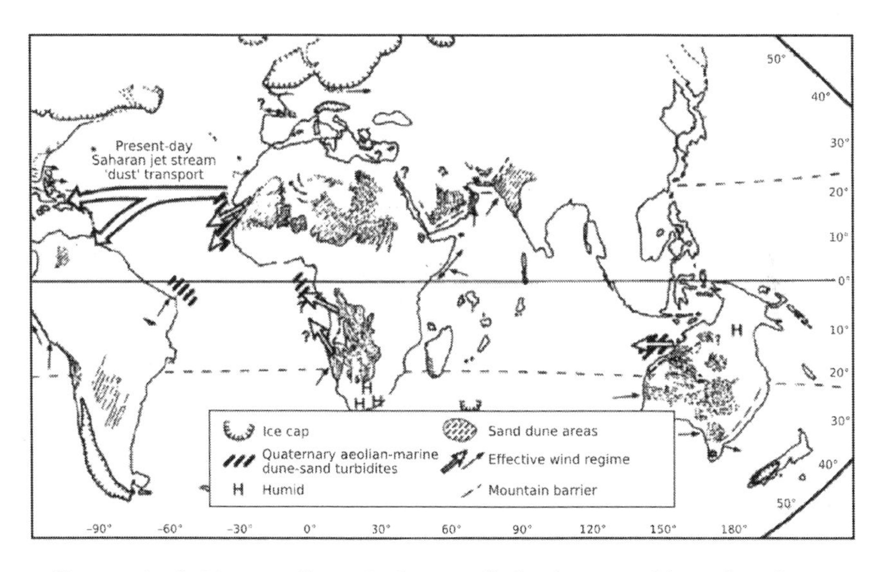

colleague Andy Vogan calls it. Gathering all the data possible within the time available is a basic prerequisite to doing your best sleuthing.

Another element we use in prospecting is the focus on 'not just what is, but what might be' (to quote Keith Williams, circa 1990). Our practice is limited by well control and perhaps seismic or other data. And certainly there is usually little data exactly where we want to drill. Thus we are always working with inferring 'what might be', between data points. That is the prize.

References

Davies, G R (1997). Aeolian sedimentation and bypass, Triassic of western Canada. *Bulletin of Canadian Petroleum Geology* **45** (4), 624–642.

The figure is from Davies (1997) and includes elements from: Sarnthein, M and L Diester-Haass (1977): Eolian sand turbidites. *Journal of Sedimentary Petrology* **47**, 868–890. It is ©SEPM.

What is shale?

Matt Hall

Until about 2007, being overly fascinated with shale was regarded as a little, well, unconventional. Seals and source rocks were interesting, sure, but always took a back seat to reservoir characterization. But in many plays today shale *is* the reservoir. And sometimes you need to know how to define something, because it affects how it is perceived, explored for, developed, and even regulated.

The Alberta *Oil and Gas Conservation Regulations* define shale as:

> *A lithostratigraphic unit having less than 50% by weight organic matter, with less than 10% of the sedimentary clasts having a grain size greater than 62.5 μm and more than 10% […] having a grain size less than 4 μm.*
>
> Section 1.020(2)(27.1), as specified in ERCB Bulletin 2009-23

This seems strict enough, but it is ambiguous. '10% of the sedimentary clasts' might be a very small volumetric component of the rock, if those 'clasts' are small enough. I am sure they meant to write '…10% of the bulk rock volume'.

The Wikipedia entry for shale cites Blatt and Tracy (1965), giving a definition that also centers on grain size, but requiring the rock to be *fissile*:

> *A fine-grained clastic sedimentary rock composed of mud that is a mix of flakes of clay minerals and [silt-sized particles]… The ratio of clay to other minerals is variable. Shale is characterized by breaks along thin laminae or parallel layering […] called fissility. Mudstones, on the other hand, are similar in composition but do not show the fissility.*

Arguably unreliable for technical definitions, dictionaries nonetheless try to describe how words are used and understood by most people. And there are specialist dictionaries, too. Here are some definitions:

> *A fissile rock that is formed by the consolidation of clay, mud, or silt, has a finely stratified or laminated structure, and is composed of minerals essentially unaltered since deposition.*
>
> Merriam–Webster Dictionary

> *Soft finely stratified rock that splits easily, consisting of consolidated mud or clay.*
>
> Concise Oxford Dictionary

A well-laminated argillaceous sedimentary rock [whose] fissility is related to the disposition of clay minerals [in] the rock. Shales do not form a plastic mass when wet, although they may disintegrate when immersed in water.

Penguin Dictionary of Geology

What about sedimentologists? Here's Potter et al. (2005):

Although the terms clay, mud and shale are widely recognized, their technical definitions and usage have long been troublesome and are not fully agreed upon. There are at least two reasons for this — the term clay is used both as a size and a mineral term, plus many clays, muds and shales are rich in silt-sized particles and thus span the clay–silt boundary.

Potter goes on to discuss some of the nuance here, then settles on *mud* as a generic fine-grained sediment, and *mudstone* as its lithified incarnation. But his point is clear: shale is, at best, a woolly term. This view is echoed by Merriman et al. (2003): '*Problems have arisen with the precise definition of the terms shale, slate and clay.*' Those authors also call out flakiness as the diagnostic characteristic of shale (as opposed to blockiness for mudstone). They add, 'shale, mudstone and slate are collectively referred to as mudrocks.'

It is interesting that none of the definitions mention organic matter, except to rule out coal. Many industry geologists today seem to use the word shale as shorthand for hydrocarbon plays that produce directly from source rocks, perhaps thinking of everything else as mudstone. The last thing the world needs is another definition, but we might summarize this notion as:

A tight, brittle rock composed mostly of mud-grade particles, containing substantial petroliferous kerogen.

So what is the geologist to do? The only safe thing to do is to be quite clear about what you mean when you talk or write about shale. If you aren't, there's a good chance that every member of your audience has a slightly different interpretation than the one you have in mind.

References & acknowledgments

Blatt, H and R J Tracy (1996). *Petrology: Igneous, Sedimentary and Metamorphic.* 2nd ed. Freeman, 281–292.

Merriman, R, D Highley, and D Cameron (2003). *Definition and characteristics of very-fine grained sedimentary rocks: clay, mudstone, shale and slate.* British Geological Survey, Commissioned Report CR/03/281N.

Potter, P, B Maynard, and P Depetris (2005). *Mud and mudstones.* Springer-Verlag, Berlin.

Shale gas development — Definition of shale and identification of geological strata. *Bulletin of the Energy Resources Conservation Board of Alberta,* Canada, July 2009 (**23**).

Whitten, D and J Brooks (1972). *Penguin Dictionary of Geology.* Penguin, Harmondsworth.

A longer version of this essay first appeared as a blog post in March 2011, *ageo.co/OIZcdo*

What's the use of fieldwork?

Jon Noad

It's a question that every manager will ask his staff sooner or later: what is the point of oil company geologists doing fieldwork? To answer this question I plan to take you on a virtual field trip, and to make things more interesting I have invited along some other technical specialists: a geophysicist, a petrophysicist, and a reservoir engineer.

Stepping from the vehicle we are confronted by a world-class outcrop. The first thing that strikes you is the scale, with maybe 100 metres of exposed rocks. Thin sandstone beds are stacked into thicker packages, and observing the outcrop as a whole one really gets a feel for the cyclicity of the rocks, in this case a series of coarsening-up parasequences, separated by thick shales. There is a real fractal feel to the rocks, with sedimentary features and bedding at all scales.

Turning to our specialists we ask the geophysicist what a seismic line through the outcrop would show. With a vertical resolution of about 20 m she estimates that we might see three reflectors in our 100 m of vertical section, certainly no more than four, but indicates the series of dipping clinoforms visible at the top of the outcrop might be recognizable. Four reflectors representing maybe a thousand individual beds doesn't sound like much, but her data can provide useful information where no wells have been drilled. Here she can see for herself what the seismic cannot image, and calibrate her intuition to nature — her next interpretation will be different.

Meanwhile the reservoir engineer scratches his head, and tells us that, when he scales up the three-dimensional geological model to simulate production in the subsurface, his cells are typically 50 m by 50 m by 1 m in thickness. He is somewhat shocked how much heterogeneity he is failing to model — everyone is learning something here.

Stepping closer it becomes clear that even the thinner sandstone beds are separated by individual layers of shale. The sandstone beds feature a variety of sedimentary structures, including cross-bedding, ripples, and a variety of trace fossils, and I explain how these can help us identify the sedimentary environment in which the rocks were formed, in this case a shallow marine setting. I ask the petrophysicist what might show up on log data from a well, and he gives us a fascinating summary on the potential resolution of gamma ray, sonic, and other logs. The gamma-ray log, for example, can resolve beds down to around 30 cm in thickness. He pulls out a formation micro-imager log (which uses resistivity measurements) to demonstrate how even burrows can be imaged using this tool.

Conversation turns to another reservoir that has challenged the reservoir engineer. He is helping to develop a new oil-sands lease in Alberta, but despite the sands exhibiting high porosities production has been unpredictable. We hop back into the vehicle and head north. In a few kilometres we reach the ancient shoreline, where thick estuarine deposits outcrop. Sandy tidal flat deposits are cut by a series of channels, and in places the channels have amalgamated into thick, stacked sandstone intervals. Approaching the outcrop we can see that each channel sandbody is cut by numerous thin mudstone beds.

I explain that these muds represent ebb tidal deposits, and that where they form baffles to flow, this can significantly affect production. The good news is that we can measure them in nature and model the subsurface better. A lightbulb goes off above the engineer's head, and he starts taking photographs — something to show the boss when he gets back to the office. Here is a perfect analog to his troublesome reservoir, and the outcrop even shows some oil staining. We collect a few samples of sandstone to take back to the lab for analysis, to compare with core samples from the reservoir itself, a kilometre below the surface.

It is abundantly clear from our virtual trip that every outcrop provides a natural laboratory to explore rock geometries and properties. In addition the rocks are a classroom in which everyone from the geologist to the technical specialist can learn about each others' disciplines. So I suggest that you get yourself out there and start observing the rocks.

Why mountains matter

Tannis McCartney

The mountains matter. Sure, everyone knows they are awesome. What geologist doesn't love going to the mountains to visit outcrops or for recreation? What I'm talking about is how the mountains matter to people studying sedimentary basins. If you're a petroleum geologist working in a foreland basin, I hope you already know that they are a major control on the geometry of the basin. Orogenic events such as continental collision, terrane accretion, and thrust faulting control lithospheric flexure and basin geometry. As petroleum geologists, we cannot ignore one of the major controls on accommodation space.

Accommodation space isn't the only reason mountains matter to petroleum geologists. Basin-wide unconformities can be tied to orogenic events. Such unconformities are major sequence boundaries, but we should also ask what conditions led to the unconformity in the rock record.

Take the Western Canada Sedimentary Basin, for example. The Jurassic sequence in the western part of the basin, which was a foreland basin at that time, is bounded by two major angular unconformities. Whatever tectonic event is causing this kind of basin-wide tilting may also be reactivating structures within the basin — structures that could be in your pool. Isn't the orogeny worth at least a second thought?

We need to think about the processes involved as interconnected gears, like in a clock. In a foreland basin, the mountains are one of the big gears.

It's not just in foreland basins that regional tectonics matter. It's not just syn-depositional tectonics that matter either. Basement matters. Pre-existing structures control the orientation of new structures. They can be reactivated during extension or compression. Carbonate reef trends can be tied to basement lineaments. Facies transitions can be controlled by basement domains.

Let's back this out one step further, and talk not just about sedimentary basins, but about all geology. It all matters. It's all connected, and while a large majority of petroleum geologists are focused on sedimentary and/or structural geology, every other branch of earth science matters too. What insights could someone studying diamondoids in heavy oil give to someone studying diamond formation in kimberlites, or vice versa?

Introductory geology textbooks like to demonstrate concepts in neat cycles like the water cycle, the rock cycle, and the Wilson cycle. As geologists we know it is more complex than this, and yet we find ourselves discretizing the cycle, only concerning ourselves with the stage that contains our rocks. When we think of a temporal sequence we need to think about the processes involved as interconnected gears, like in a clock. In a foreland basin, the mountains are one of the big gears.

Acknowledgments

The photograph is licensed CC-BY by Flickr user Sam Beebe.

List of contributors

Nigel Banks has more than 30 years of experience as a geoscientist in petroleum exploration and production. He worked initially in international exploration for Shell and Occidental before taking up management positions with UK consulting companies. He now works as an independent consultant. He is also a visiting lecturer at Imperial College, London where he teaches on the petroleum MSc programs.

Jim Barclay is the Geologic Advisor in the Western Canada Business Unit at ConocoPhillips Canada in Calgary, Alberta. He is a graduate of Carleton University and the University of Calgary, and after a brief fling with mineral exploration, started his petroleum career at Dome Petroleum. Jim was Communications Director of the Canadian Society of Petroleum Geologists.

Evan Bianco is the chief scientific officer at Agile Geoscience. He is a blogger, entrepreneur, and knowledge-sharing aficionado. He has an MSc in geophysics from the University of Alberta and five years' experience as an industry consultant in Halifax, Nova Scotia. Evan tries to teach himself something new every day, and every so often, it proves useful. He can be reached at *evan@agilegeoscience.com*, or you can follow him on Twitter *@EvanBianco*.

Clare Bond is a structural geologist at the University of Aberdeen, UK. She graduated from the University of Leeds and completed a PhD at Edinburgh University. Clare then worked in geological conservation and policy roles before taking a position at Midland Valley Exploration, consulting worldwide and leading their knowledge centre and R&D initiatives. She enjoys interdisciplinary research and has collaborated with social scientists, psychologists, and modellers. She can be reached at *clare.bond@abdn.ac.uk* and visited on the web at *ageo.co/clarebond*.

Stuart Burley is head of geoscience at Cairn India, and is based in Delhi. After graduating with a PhD in geochemistry from the University of Hull, Stuart enjoyed a 15-year career in academia — at Berne, then Sheffield, and finally Manchester University. In 1995 he embarked on a career in industry, spending almost 13 years at BG Group in positions all over the world. He has been at Cairn since 2008.

Alex Cullum is a geologist, palaeontologist, and author, currently working within exploration for Statoil in Stavanger, Norway. Psychology and how we communicate science and the things that matter are the subject of his new novel.

Mark Dahl is a geologist for ConocoPhillips, currently working in Lower 48 New Ventures in Houston, Texas. He has a degree in geography from the University of Victoria, Canada (2003), and started his career in GIS and CAD, before switching to geology and graduating with a degree in geology from the University of Calgary in 2008.

Shahin Dashtgard is an Associate Professor of Geology at Simon Fraser University. He graduated from the University of Alberta in 1998 and worked at Fletcher Challenge and Talisman before embarking on a PhD in sedimentology. He moved to Simon Fraser in 2007 after a brief stint at the Alberta Geological Survey. Shahin is an Associate Editor of the *Journal of Sedimentary Research* (SEPM) and of *Ichnos* (International Ichnological Association), and was the recipient of the SEPM James Lee Wilson Award in 2012.

Paul de Groot is president of dGB Earth Sciences, *dgbes.com*. He worked 10 years for Shell where he served in various technical and management positions. Paul subsequently worked four years as a research geophysicist for TNO Institute of Applied Geosciences before co-founding dGB in 1995. Paul has authored many papers and together with Fred Aminzadeh, wrote a book on soft computing techniques in the oil industry. He holds MSc and PhD degrees in geophysics from Delft University of Technology in the Netherlands. Find him on LinkedIn at *ageo.co/IhwEUg*.

Clayton Deutsch is Director and Professor in the School of Mining and Petroleum Engineering, in the engineering department at the University of Alberta. He teaches and conducts research into better ways to model heterogeneity and uncertainty in petroleum reservoirs and mineral deposits. Prior to joining the University of Alberta, Clayton was an associate professor in the Department of Petroleum Engineering at Stanford University. He has also worked for Exxon Production Research and Placer Dome Inc. He has published six books and over 200 research papers, and holds the Alberta Chamber of Resources Industry Chair in Mining Engineering and the Canada Research Chair in Natural Resources Uncertainty Characterization.

Tony Doré is a senior advisor at Statoil with a global exploration remit. He lives in London. He was previously VP of Exploration North America and VP of Exploration New Ventures North America. Tony has published more than 50 peer-reviewed papers, and edited six books. He has served academia and industry in various capacities, and in 2010 was appointed as an Officer of the Most Excellent Order of the British Empire (OBE).

Jesper Dramsch is a geophysicist at heart. At the age of 17 he was the first student to be granted access to college courses in geophysics as part of a junior studies program at the University of Hamburg. He then pursued a BSc in geophysics and oceanography. His bachelor's thesis in seismic interpolation was published as an extended abstract by the EAGE. He has worked as a geophysical intern with Fugro Seismic Imaging (now CGGV) and Schlumberger, and is currently working on his Master's thesis on pre-stack data enhancement for improving sub-salt illumination. He blogs about diverse geoscientific and related topics that catch his interest at *dramsch.net* and he is *@JesperDramsch* on Twitter.

Alan Gibbs is the founder of Midland Valley, a UK software and consulting company. After leading the company for 30 years he is now chairman advising on global leadership and strategy in the company. Alan previously worked as an academic and as a senior specialist in a state company. He is responsible for a number of seminal publications in the application of structural geology to oil and gas exploration and production. Reach him at *alan@mve.com.*

Matt Hall is the founder of Agile Geoscience. He graduated with a PhD in sedimentology from the University of Manchester, UK, in 1997. Matt has worked at Statoil in Stavanger, Norway, Landmark and ConocoPhillips in Calgary, Alberta, and is now running his consultancy firm Agile from the beautiful town of Mahone Bay, Nova Scotia. He is passionate about communicating science and technology, and especially about putting specialist knowledge into the hands of anyone who needs it. Find Matt on Twitter as *@kwinkunks* or by email at *matt@agilegeoscience.com.*

Richard Hardman was exploration director of Amerada Hess at a time when the company was the UK's third largest oil producer. He was president of the Geological Society of London from 1996 to 1998, and of the Petroleum Exploration Society of Great Britain. He was awarded the William Smith medal of the Geological Society in 2003, and was appointed Commander of the Most Excellent Order of the British Empire (CBE) in 1998 for services to the oil industry.

John Harper is a researcher in industry, academia, and government. He was Professor of Petroleum Geology and Sedimentology at Memorial University of Newfoundland, and has worked at Shell Development, Shell Oil, Shell Canada, Trend Exploration, ConocoPhillips Canada, the Geological Survey of Canada, and EBA Consulting Engineers. John's greatest strengths are in integrated geological and geophysical basin and reservoir characterization, structural and depositional system analysis, and training and mentoring.

Bruce Hart is currently a research geologist at Statoil in Houston, Texas. He previously worked for ConocoPhillips, McGill University, New Mexico Tech, Penn State, and the Geological Survey of Canada. He has taught seismic interpretation courses since 1995 in Houston, Calgary, London, Cairo, Copenhagen, Kuala Lumpur, Vienna, Denver, and elsewhere.

Dan Hodge graduated with a BSc in geology from the University of Manchester in 1997 and an MSc in geology from Royal Holloway, University of London, in 2001. He worked at Midland Valley in the UK before moving to Canada to work for Talisman. He now works at Parex Resources in Calgary, Alberta.

Nicholas Holgate completed his MSci degree in geology at the University of Bristol, UK, in 2006, before working as an exploration geologist for Resolve Geological in Australia. He left Resolve Geological to undertake a PhD at Imperial College, London, supervised by Chris Jackson and Gary Hampson.

Duncan Irving is an oil and gas practice lead at Teradata, a global provider of analytic data platforms. He wants to revolutionize the way subsurface and reservoir operations are driven by the smooth flow of data and decision making, and is proud to be building massive-scale computational platforms to enable this. Duncan graduated in geophysics from the University of Edinburgh in 1995, and earned a PhD in glacial geophysics from Cardiff University/ETH Zurich in 1999.

Chris Jackson completed his BSc (1998) and PhD (2002) at the University of Manchester, UK, before working as a seismic interpreter for Norsk Hydro in Bergen, Norway. He left Norsk Hydro in 2004 to take an academic position at Imperial College, London, where he is now Reader in Basin Analysis. In 2013 Chris was an AAPG Distinguished Lecturer.

Zane Jobe graduated with a PhD in sedimentology from Stanford University in 2010, with a thesis on channelized turbidite systems. He is now a research geologist at Shell in Houston. Zane is *@ZaneJobe* on Twitter and writes a blog at *offtheshelfedge.wordpress.com*.

Fraser Keppie is an earth scientist with an interest in the kinematic and geodynamic evolution of orogenic belts and sedimentary basins. In particular, Fraser is interested in developing predictive models of paleogeography for applications as diverse as understanding mantle–lithosphere processes at depth, climate–lithosphere processes at surface, and exploration models for mineral and energy resources in the shallow crust.

Derik Kleibacker is currently the chief geologist for ConocoPhillips Indonesia. He graduated with an MSc degree in geology from Oregon State University and has worked for ConocoPhillips in various exploration and development roles since 2002. Derik hates to admit it, but he loves geophysical interpretation as much as geological field mapping.

Elisabeth Kosters is a geological consultant based in Wolfville, Nova Scotia. She graduated from the University of Amsterdam with a master's degree in Quaternary geology and from Louisiana State University with a PhD in sedimentology. She has more than 30 years experience in government and academia, and as an independent consultant. Most recently, Elisabeth served as Executive Director of the Canadian Federation of Earth Sciences. She was president of the Atlantic Geoscience Society and of the Royal Netherlands Geological and Mining Society.

Erik Lundin has a geology degree from Fort Lewis College, Durango, and a masters in geology from the University of Arizona, Tucson. He has a Dr Philos from the University of Oslo. He works as an exploration geologist for Statoil in Norway. His professional interests are tectonics and regional geology, with a focus mainly on the North Atlantic and Arctic.

Aruna Mannie holds a BSc in petroleum geoscience from the University of the West Indies and an MSc in petroleum geology from the University of Aberdeen. Her favourite research projects are those that involve prospect generation and volumetric calculations. She spent six years of her professional career as an exploration geoscientist working for BG Group and Petro-Canada. Aruna is currently pursuing a PhD at Imperial College on the influence of rifting on salt tectonics and shallow-marine sedimentation.

Allard Martinius is interested in clastic facies analysis, sedimentology, and sequence stratigraphy of fluvial and shallow marine deposits, as well as in clastic reservoir characterisation and geomodelling. He has been team lead on long-term research projects at Statoil's research centre in Norway. Allard has experience from the Norwegian continental shelf and heavy-oil development projects in Venezuela and Canada, and has worked on numerous other projects around the world.

Tannis McCartney completed her BSc in geophysics at the University of Alberta in 2000. After spending six years working for geophysical service companies in Calgary, she returned to university for graduate work. She completed an MSc in geology at the University of Calgary in 2012 and is now at Syracuse University pursuing a PhD in sedimentation and tectonics in the Lacustrine Rift Basin Research Group, studying basin evolution in Lake Malawi. She is *@TMMCC* on Twitter and blogs at *tannislikesrocks.blogspot.ca.*

Tom Moslow has a PhD in geology from the University of South Carolina and an MS in geology from Duke University. He is an Adjunct Professor in the Department of Geosciences at the University of Calgary and a former Professor at the University of Alberta (1985–1994) and Louisiana State University (1983–87). Tom worked at Canadian Hunter Exploration and Ulster Petroleum before co-founding Midnight Oil and Gas Ltd in 2000 where he served as Vice President of Exploration. This led to the creation of Daylight Energy Trust where he served as Vice President of Geology and Geophysics. He held executive positions in subsequent affiliated companies including Midnight Oil Exploration and Pace Oil and Gas Ltd. He retired from Pace at the end of 2011 and formed Moslow Geoscience Consulting. Tom has won numerous awards and honours and has taught a variety of professional development courses on applied petroleum geology subjects. He has authored or coauthored over 75 publications in a variety of journals.

Mark Myers is former State Geologist and head of Alaska's Geological Survey, and Director of the United States Geological Survey from 2006 to 2009. Mark received his doctorate in geology from the University of Alaska–Fairbanks specializing in sedimentology, clastic depositional environments, surface and subsurface sequence analysis, and sandstone petrography. He earned his bachelor's and master's degrees in geology from the University of Wisconsin–Madison. He served as an officer in the US Air Force Reserve from 1977 to 2003, retiring as lieutenant colonel .

Jon Noad graduated from Imperial College, London in 1985, and started work as a mining geologist on gold mines in South Africa. He moved to the Bushveld in 1988 to work on a shaft-sinking platinum mine, then returned to the UK in 1990 and joined the telecommunications industry. While there, he completed an MSc in sedimentology at night school. Jon then returned to the University of London full time to complete a PhD in Borneo, before being recruited by Shell in 1998, eventually moving to Shell Canada in 2006. He worked western Canada exploration for four years and then joined Murphy Oil as Exploration Manager, Canada in 2010. He left Murphy in 2012 to join Husky Energy where he works as Geological Specialist. He is never happier than when in the field sharing geology with others.

S George Pemberton received his PhD from McMaster University in 1979. He is currently a Distinguished University Professor and the C R Stelck Chair in Petroleum Geology in the Department of Earth and Atmospheric Sciences at the University of Alberta. The main thrust of his research pertains to the application of ichnology to petroleum exploration and exploitation. His work has been recognized by a number of awards including: the GAC Past President's Medal (1994); the 2003 R C Moore Medal for Excellence in Paleontology, presented by SEPM; the 2008 Grover Murray Distinguished Educator Award presented by AAPG; the 2009 Killiam Award for Excellence in Mentoring; and the 2013 Logan Medal from the Geological Association of Canada. He was elected a Fellow of the Royal Society of Canada in 2001; Canada Research Chair 2002–09 in Petroleum Geology; and honorary membership in CSPG in 2010. In addition he has won 15 best paper or best presentation awards, has published over 225 papers, and edited or co-edited seven books.

Gary Prost graduated from Colorado School of Mines with a PhD in geological engineering in 1986. He is a geologist in the Oil Sands division at ConocoPhillips Canada, and has worked on plays all over the world. His book *Remote Sensing for Geoscientists: Image Analysis and Integration* published by CRC Press, is now in its third edition.

Michael Pyrcz is a research scientist with Chevron. His education includes a BSc in mining engineering and a PhD in geostatistics from the University of Alberta. He has completed over 25 peer review publications, numerous conference talks and invited talks, and taught various short courses. Recently he completed the second edition of *Geostatistical Reservoir Modeling* (Pyrcz and Deutsch 2014), a geostatistics textbook for Oxford University Press. Michael's recent research interests in reservoir geostatistics include methods for improved geological realism and uncertainty models for exploration.

Heterogeneity + sparse sampling = uncertainty 34

Tim Redfield is a structural geologist based at the Norwegian Geological Survey in Trondheim, Norway. He graduated from the University of California–Santa Cruz in 1985, completed his master's at Western Washington University in 1987, and took his PhD from Arizona State University in 1994.

The topography of tectonics 92

Weishan Ren is a principal geologist and geomodeller at Statoil in Calgary, Alberta. He is a geomodelling specialist. Weishan graduated with a PhD from the University of Alberta in 2007, having studied in Clayton Deutsch's research group, and worked at ConocoPhillips Canada until joining Statoil in 2011.

Three kinds of uncertainty 98

Brian Romans is a sedimentary geologist and assistant professor in the Department of Geosciences at Virginia Tech. He graduated from SUNY Buffalo with a geology degree in 1997 and then worked for small oil and gas companies in Buffalo, New York and Denver, Colorado. Brian received an MS in geology from Colorado School of Mines in 2003 and a PhD in geological and environmental sciences from Stanford University in 2008. He worked as a research geologist for Chevron for a few years before joining the faculty at Virginia Tech in 2011. Learn more about Brian's research on sedimentary systems and download his published papers at *www.geos.vt.edu/people/romans*. Brian is *@clasticdetritus* on Twitter and writes the blog *Clastic Detritus*.

Strata are not flat 76

Bernd Ruehlicke is president of Eriksfiord Inc. part of the Eriksfiord group, a geoscience provider with a focus on image and sonic-log based geological and geomechanical studies. Bernd contributed to most of the geological applications modules of Recall (now Halliburton/Landmark) during his time at Z&S in Stavanger, Norway. At PGS and Landmark, he built the interface between the Petrobank (Oracle) database and Recall, and worked on R&D projects such as the Java DecisionSpace platform. Bernd is now developing the business of the Eriksfiord group in the Americas. He holds a BS in computer science and MS in mathematics from Aarhus University, Denmark. He has an MBA from the University of Houston–Victoria.

Get to know eigenvectors 30

Zoltán Sylvester became interested in deepwater sediments because turbidites make up most of the mountains surrounding his Transylvanian hometown; and started learning about geology at Babeş-Bolyai University, in Cluj, Romania. He went on to study turbidites at Stanford University, where he received a PhD in 2001. Since then he has been working as a clastic sedimentologist and seismic stratigrapher at Shell's research centre in Houston, and is interested in figuring out how to describe, represent, and predict the complexities of stratigraphy, from the scale of sand grains to that of seismic volumes. Zoltán is *@zzsylvester* on Twitter and blogs at *hinderedsettling.com*.

Dianne Tompkins holds a BSc degree from Aston University, UK, and a PhD from Imperial College, University of London. She has worked all over the world, and is currently a consultant geologist in Stavanger, Norway. Dianne has been active in AAPG since 1990.

Patrick Walsh is chief geologist for Ormat Technologies Inc and oversees the technical aspects of the company's western US and international projects. He is also the lead geologist for several Ormat projects including Steamboat Hills, Glass Buttes, and Brawley. With 13 years of experience, Walsh has specialized in geothermal, water, and petroleum resource assessment, exploration, and development.

GEOGRAPHICAL INDEX

Page numbers of places mentioned in the text.

Index

AGILE LIBRE
52 THINGS YOU SHOULD KNOW ABOUT
GEOPHYSICS
EDITED BY MATT HALL & EVAN BIANCO

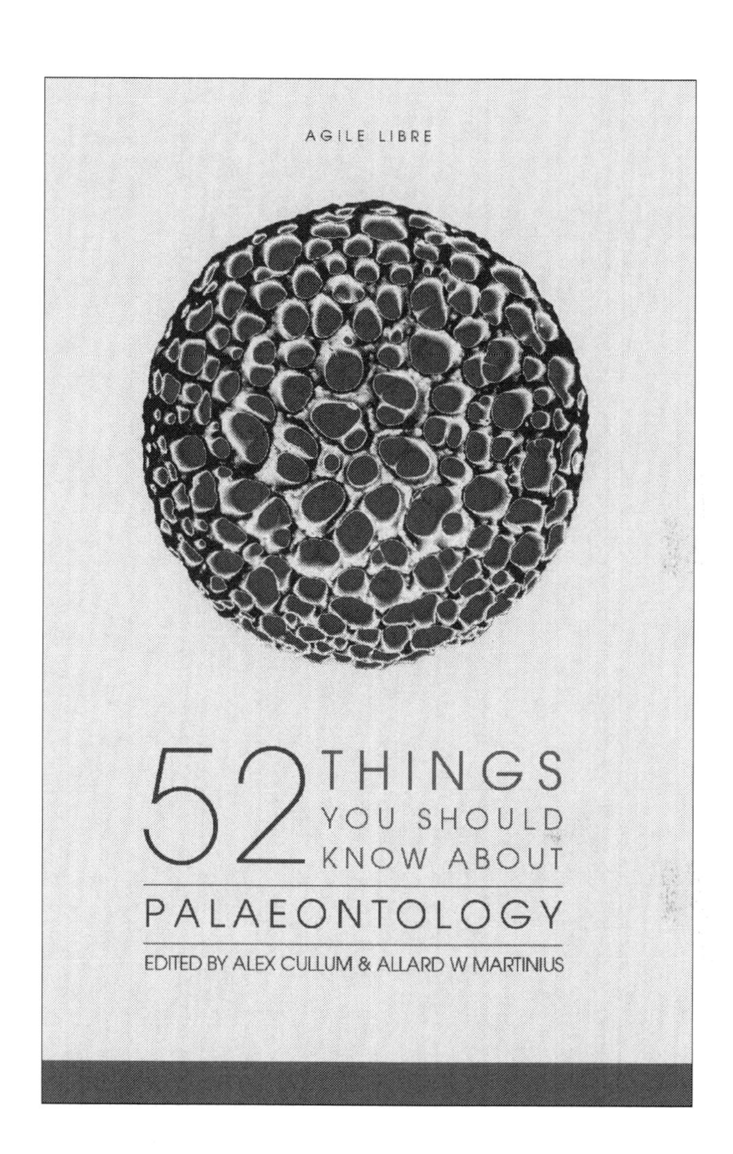

Available from ***ageo.co/52palaeo*** and
online bookstores worldwide.

ISBN 978-0-9879594-4-7

41850932R00076

Made in the USA
San Bernardino, CA
21 November 2016